Lecture Notes in Mathematics

A collection of informal reports and seminars
Edited by A. Dold, Heidelberg and B. Eckmann, Zürich

Series: Mathematisches Institut der Universität Heidelberg
Adviser: K. Krickeberg

120
Decidable Theories I

Edited by Gert H. Müller, Heidelberg

9783662358481

Dirk Siefkes

Mathematisches Institut der Universität, Heidelberg

Büchi's Monadic Second Order Successor Arithmetic

Springer-Verlag
Berlin · Heidelberg · New York 1970

To Marie Luise

Editor's Preface

For several years I conducted a Seminar on decision procedures of various mathematical theories at the Mathematical Institute of the University of Heidelberg, some of the time jointly with Professor O. Herrmann. It is planned that I should continue this seminar with Dr.D.Siefkes. The results will be published in a subseries of "Lecture Notes in Mathematics", to be called "Decidable Theories". The following aspects were, and will be, decisive both for the seminar and the Lecture Notes arising out of it.

(i) The decidable theory in question and the decision procedure are set up in purely syntactical terms, hence not referring, for instance, to the true sentences of a preferred model. The transformation steps which in the simplest case lead to an equivalence with the propositional constants T (true) or F (false) are proved within the theory in question, - of course, using syntactical means.

(ii) A presentation of a decision procedure in the sense of (i), provides a set of operational devices which, if applied to a sentence, leads to its decision. From such devices certain syntactical parameters of a sentence can be exhibited, and in these terms an estimation of the complexity of the decision procedure can be made. - Obviously, most presentations of decision procedures in the literature are given so that the decidability is proved in the simplest mathematical way. Hence it is not surprising that a presentation aiming at practical applications results in some remarkable simplifications of the decision procedure.

(iii) A decision procedure usually is: 1) the reduction of an arbitrary sentence to a normal form (using the axioms of the theory) and 2) the reduction of a given sentence in normal form to T or F, as usual. - Relative to a preferred model, e.g. interesting from the point of view of applications, the normal form indicates the type of question which can be answered by the use of the decision procedure. There may be different normal forms for one theory, to which a suitable reduction can be applied; hence different types of questions may be decidable. It is of interest to find questions in science or technology which are reducible to such types.

The aim of the Lecture Notes on Decidable Theories is a systematic and almost self-contained presentation of the known decision procedures, taking into account (i)-(iii). It should in principle be possible to program a decision procedure from such a presentation. Substantial simplifications may arise if certain syntactical parameters are numerically given in advance. - It is the intention of the Series to bridge the gap between applications and the more or less abstract results on decidability.

Heidelberg, January 6, 1970   Gert H. Müller

C O N T E N T S

Introduction

Introduction

The subject of these lecture notes is Büchi's decision procedure (see [3]) for his Sequential Calculus SC, i.e. for the monadic second order fragment of arithmetic which involves only the successor function, but neither addition nor multiplication. We set up a decision procedure for SC in purely syntactical terms, using none but the three Peano axioms for successor. As a result we prove the Post-completeness (syntactical completeness) of this system, and obtain several measures towards practical effectivity of the decision procedure (chap.I). In chap.II we investigate definability in SC. We introduce three extensions of the concept "ultimately periodic" from sets of natural numbers to relations, and prove that one of them is appropriate for SC, namely that just the "fanlike ultimately periodic" relations are definable in SC. This result provides especially a characterization of the functions definable in SC, and yields effective quantifier elimination for SC-formulae without free predicate variables. (Chap.II.2.)

In a monadic second order theory one has at hand quantifiable one-place predicate variables, besides the means of first order logic. During the last ten years monadic second order theories have turned out to be a very powerful tool in establishing decidability. The bound predicate variables allow to express a lot of combinatorial facts within the theory; e.g. all known decision procedures for monadic second order theories use finite automata on finite or infinite tapes or trees, by translating automata theory into the investigated theory. The engagement between automata and monadic second order started in the mid-fifties from several papers of Church who uses quantifier-free fragments of monadic second order arithmetic as "condition languages" for automata theory (see [8]-[10]). The marriage was performed by Büchi; in his papers [2]-[5] (see also [6] and [7]) for the first time the decidability of several monadic second order theories is shown, and for the first time finite automata on infinite tapes are used for such proofs. It is easily seen, and is done in the author's Diplom paper, that all known decision procedures (and also a lot of new ones) for fragments of arithmetic can be derived from Büchi's decision procedure [3] for SC. How fruitful the mariage was, became even more evident in the papers of Elgot-Rabin [12] and Rabin [28] and [29]. In [28], Rabin solves a lot of famous decision problems in the affirmative by proving the decidability of the monadic second order theory of two (or of more) successors; his main tool is the theory of automata on infinite trees. - Thus it seems worthwhile to investigate decidable monadic second order theories. Especially, SC calls for a closer inspection: on the one hand, most of the consepts useful in monadic second order theories are introduced already in [3]; on the other hand, Büchi's decision procedure is published only as a very short congress talk paper which leaves a lot of work to the reader.

A decision procedure for a theory is a mechanical procedure which, applied to any sentence of the theory, tells after finitely many steps whether the sentence is true or not. In most cases, a decision procedure is given semantically, i.e. the true sentences of the theory are prede-scribed somehow, and then model-theoretical methods are used to show how to transform any sentence of the theory step by step into an "evidently" true or false one. To get a syntactical version of the decision proce-dure one has to write down within the language of the theory all trans-formations of formulae which appear in these steps; especially one has to translate all means from outside into principles which are expressible within the used language. The thus collected transformations constitute an axiom system from which any true sentence of the theory is derivable. Therefore the best characterization of the theory would be a simple part of this axiom system from which the remaining transformations are deri-vable. – Such a "syntactization" indirectly gives a further account of the strongness or weakness of the theory. Namely it shows what theorems are derivable or not derivable in the theory. In general, it leads to certain normal forms for formulae, and thus marks off the "range of questions" of the theory; i.e. one sees more clearly what sort of prob-lems can be formulated in the theory.

In chap.I, this program of syntactization is carried through for the Sequential Calculus SC. Büchi sets up semantically both his system and the decision procedure. As means from outside he uses results from the theory of finite automata, and the famous combinatorial theorem A of Ramsey [31]. (It should perhaps be pointed out that by syntactization one does not get rid of these means from outside. Syntactization just makes the proofs "elementary", i.e. expressible in the language, and thereby shows the degree of complexitiy of the theory. But the elemen-tarized proofs mostly are very cumbersome, and not understandable but from their intuitive (= outside) formulation.)

The main points of chap. I are:

(i) In §§1-3 a syntactical decision procedure for SC is presented – for the first time, as far as we know. Clearly the procedure follows Büchi's semantical procedure, in the manner described above.

(ii) It is shown that on the background of monadic second order logic from the three Peano axioms for the successor function, recursion theory as far as it is expressible can be built up in the system. This part of recursion theory suffices to derive the Ramsey theorem, suggested by Büchi as an axiom, and to replace automata theory. At one place, recursion formulae, which work like finite automata in the system, allow to avoid at all the most involved contribution from automata theory. Thus it is shown that this very simple axiom system is complete. Derivation of recursion theory and of the Ramsey theorem in SC are to be found already in [36] - however, in a more complicated version than here.

(iii) In §4 the decision procedure is inquired with regard to feasibility. - A step-by-step-description of the whole procedure is given, to encourage people interested in application but not in proofs, to try to program the procedure on a computer. Estimations of the growing of the length of formulae under the transformations of the decision procedure indicate the most awful parts of the decision procedure (awful with regard to practical application), and give reason to quite a lot of improvements. E.g. we replace at a central place of the decision procedure certain formulae used by Büchi by simpler ones, and lower thereby a growing rate of 2^{2^a} to 2a. In 5.c, it is just the most intrinsic part of the completeness proof, which suggests how to replace another part of the decision procedure by a simple combinatorial consideration. These improvements together make it more likely that practical application of the decision procedure is possible.

(iv) Most effort is made to make clear the intuitive background of the decision procedure. Tuples of predicates are regarded as sequences of tuples of truth values, and SC-formulae containing free predicate variables are regarded as conditions on such sequences, called "threads". The research on nerve nets once has led to the theory of finite automata as a formal tool of analysis. Threads directed by conditions on the other hand, seem to be the best informal means for the neuro-biologist to formulate his problems, and then to undertake the investigation by formal means somehow. By thinking in directed threads, moreover, one gets a clear picture of the meaning of the normal form $\Sigma_\wedge^\omega$ for SC-formulae, which is the main objective of the decidability proof, and which is itself relatively easy to decide. The investigation of $\Sigma_\wedge^\omega$-formulae shows that just the ultimately periodic predicates are definable in SC; thus SC-conditions are good to determine ultimately periodic threads, but no others. In this way the "range of questions" is described, and thus it may be that neuro-biologists and others will be able to translate a good part of their statements into $\Sigma_\wedge^\omega$-form, and therefore to make good use of the decision procedure in avoiding its most terrifying part.

From a decision procedure mostly follows an account of definability, i.e. a description of the sets and relations, especially functions, definable in models of the theory considered. Such a description is not only of theoretical interest, but gives new information, namely: (a) Information on the formal theory itself; e.g. one can get new normal forms by showing equivalence of classes of formulae. (b) Information on the mathematical content of the theory; e.g. one can get a classification of the models of the theory. It can be very interesting to compare this classification with classical results on the theory, i.e. with the "mathematical" theory not restricted to a fixed logic calculus. (c) Information in turn on the decision procedure; it is easier to understand the steps if one has a concrete imagination of what the particular formulae mean in the model. (d) If the decision procedure is not already effected by a method of quantifier elimination, it may be that one gets conversely such a method by the knowledge on definability.

Investigation of definability along these lines is the main content of chap.II. In II.1.a, following Büchi [2],[3] and Church [8]-[10], it is shown that the restricted recursion formulae of Church, our recursive SC-formulae, and finite automata, all are equivalent in defining sets of words (where a word is a finite part of a thread). - Büchi has shown that just the ultimately periodic sets of natural numbers are definable in SC. We define in §2 the new concept of "fanlike ultimately periodic" relations over natural numbers, and show that just these relations are definable in both SC and the elementary theory CO of congruence and order. It follows that e.g. among the monotonic increasing functions exactly those are definable which are ultimately either constant or a periodic deformation of the identity function. (Thereby the statement of Büchi [2],[3] that any linear function is definable in SC, is corrected.) Further we obtain an effective procedure to eliminate quantifiers from SC-formulae without free predicate variables. Moreover, we prove that one gets a model for SC if one allows instead of arbitrary sets of natural numbers only ultimately periodic sets as interpretations for the predicate variables. At last, in II.3.c we show how to translate the decision procedure for SC from the natural numbers to the integers.

These lecture notes and further results, e.g. on decidable and undecidable extensions of SC (see [35] and [37]),are very much influenced by seminars on "Decidable theories" which were held at Heidelberg by Professor Gert H. Müller and Professor O. Herrmann through several years. I would like to express my thanks to all participants of these seminars. But my deepest thanks are due to Professor Müller himself: most parts of this paper were stimulated by his ideas, and he never hesitated to spend his time to discuss the resulting problems. Especially

it was his suggestion to eliminate the use of automata from the decision procedure to make a completeness proof possible; conversely he always insited on the need for smooth concepts to make better understandable both the fact and the proof of the decidability, and for simplification in the procedure to make it applicable. But as a very fact,
after so many discussions it is impossible to untie his influence and
my own work. I express a special thank to my wife; without her untiring
effort in reading and typing so many manuscripts these lecture notes
would have never appeared.

Heidelberg, October 27, 1969   Dirk Siefkes

Technical Hints for the Reader

Numbers in square brackets refer to the bibliography at the end of
the paper. - Instead of making use of page numbers, we refer to other
parts of this paper with the help of chapter, § and section. Thus
"theorem I.3.b.3" means "theorem 3 of section b of §3 of chap.I"; if
the result referred to is to be found in the same chapter, or §, or
section, we delete the corresponding initial part of the code word,
thus writing e.g.: lemma 5.c.1, corollary b.2, definition 4. In the
same way we refer to sections: I.3.b, 3.b, b. - Throughout the paper we
write "DP" and "iff" short for resp. "decision procedure" and "if and
only if". Further we use the set-theoretical symbols $\cup, \cap, \{\}, \pi, \subset$ to denote resp. union and intersection of sets, set abstraction, power set,
and set inclusion. To enhance clearness we indicate the end of a proof
by the sign $\#$. - For all other abbreviations and introduced notations
see the "list of symbols and notations" behind the bibliography.

Chapter I. Decidability and completeness of SC

In accordance with the plan of the introduction in this chapter we work out the DP of Büchi[3] into two directions: (i) We present a full syntactical version of the DP for the sequential calculus SC. (ii) We give a step description of the DP so that one can perform the DP consulting only this list.

Ad (i): An explicit description of the syntactical system SC will be found in 1.a. The rest of § 1 serves to explore the power of this system in trying some formal proofs: In 1.b recursion theory is built up as far as needed in the sequel. From the recursion theorem, in 1.c the theorem A of Ramsey[31] is derived. Stepwise natural simplification of formulae leads in 1.d to certain normal forms $\sum_{\mathcal{W}}^{\omega}$ for $1 \leq \mathcal{W}$, which suggest two main problems: to show that $\sum_{\mathcal{1}}^{\omega}$ is (1) closed under negation, and (2) decidable. Since only few readers would like to read proofs which consist just of sequences of formulae, we have chosen a half formal presentation: In 1.a we introduce "processes" as natural interpretation of tuples of predicate variables, and regard formulae as conditions on processes; this seems to be a good help in understanding formulae and transformations of formulae, and thus in understanding proofs. In 2.a, ω -sequences of tuples of truth values, called threads, are introduced as formal counterpart of processes. The need to consider finite pieces of threads suggests two other normal forms, $\sum^{\circ}$ and $\sum_{R}^{\circ}$, the formulae of which work as resp. nondeterministic and deterministic finite automata in the system. By the above interpretation, the decidability of $\sum_{R}^{\circ}$, $\sum^{\circ}$ and $\sum_{\mathcal{1}}^{\omega}$ is shown fairly easily in 2.b, whereby main problem (2) is solved. After the proof in 2.c that $\sum_{R}^{\circ}$ and $\sum^{\circ}$ are closed up to equivalence with respect to Boolean operations, main problem (1) — to show the same fact for $\sum_{\mathcal{1}}^{\omega}$ — is attacked in 2.d (there the Ramsey theorem is used), and solved in §3. In 5.a, the completeness of the axiom system for SC is established; a more elaborate derivation is postponed to 5.b.

Ad (ii): In 4.a we collect the scattered results to get the decidability, and present a list of the steps of the DP. An estimation of how much the length of formulae grows under performance of the DP, is given in 4.b. This estimation leads in 4.c to some improvements of the DP, which make it more likely that the DP can be performed on a computer, at least for some easier types of formulae. A little more light is thrown on the DP by consideration of two examples in 4.d. At last, the derivation quoted at the end of the last paragraph suggests in 5.c a further simplification of the DP and therefore of the completeness proof itself.

§1. The sequential calculus SC

a) The system

Monadic second order theories - the object of this paper - are formalized within monadic second order predicate calculus as are elementary theories within equality calculus: in both cases the calculus is extended by non-logical constants and by axioms. Representatively for the theories dealt with in this paper we give now an exact description of SC, including the logical rules and axioms. This accuracy pays: it turns out that the seemingly harmless substitution rule for predicate variables is very powerful; namely its equivalent, the comprehension principle, is used in most derivations, thus is a fundamental principle in the completeness proof.

Object language: As individual variables we use small Latin letters: a,...,e as free, t,x,y,z as bound variables. Analogously A,...,E,G,H and P,Q,R,S as resp. free and bound one-place predicate variables. All variables of the object language and the metalanguage may be indexed by natural numbers. The quantifiers $\forall$,$\exists$ serve for both types of variables. Further we use: sentential connectives $\wedge$,$\vee$,$\neg$,->,<->; T and F for the truth values "true" and "false"; square brackets $\ulcorner$,$\urcorner$,$\ulcorner$,$\urcorner$,... and dots for bracketing formulae; round brackets to include quantifiers and the arguments of predicate variables: e.g. $(\exists P)(\forall t)P(t)$. As only nonlogical signs we employ the individual constant o to denote the zero-element, and the one-place function symbol ' for the successor function. We use small Gothic letters from the middle of the alphabet $(\mathfrak{k},...,\mathfrak{y})$ to abbreviate superpositions of the successor function: a+$\mathfrak{w}$ instead of a$\underset{\mathfrak{w}}{\underbrace{'...'}}$, and simply $\mathfrak{w}$ instead of o+$\mathfrak{w}$. This notation should not be confused with an introduction of addition into SC. Terms like a+b are not available in SC; indeed, enriching by addition would make SC undecidable. To be precise: Terms like a+17 could be (but will not be) introduced into the object language by explicit definitions; but in a+$\mathfrak{w}$ the $\mathfrak{w}$ is a variable of the metalanguage helping to indicate such terms. - Later on we shall introduce by explicit definitions the equality sign =, further < for "smaller than", $\equiv$ ($\mathfrak{w}$) for "congruent modulo $\mathfrak{w}$" ($\mathfrak{w}$ > o), and other signs. To avoid clumsy formulation we denote $\mathfrak{n}$-tuples of variables by underlining (and sometimes by an upper index $\mathfrak{w}$), e.g. $\underline{P}^{\mathfrak{w}}$,$\underline{a}^{\mathfrak{m}}$ instead of $P_1,...,P_{\mathfrak{n}}$ or $a_1,...,a_{\mathfrak{m}}$ (for example see next paragraphs); these strings of variables may be indexed by natural numbers, too. To save brackets we agree upon: (1) dots extend over brackets, (2) $\neg$,$\forall$ and $\exists$ make a lesser break than $\wedge$ and $\vee$, and $\wedge$ and $\vee$ make a lesser break than -> and <->. Dots are used mostly to indicate that quantifiers extend over the whole formula, e.g. $(\exists P).\mathfrak{A}(P) \vee \mathfrak{L}(P)$ instead of $(\exists P)\lceil\mathfrak{A}(P) \vee \mathfrak{L}(P)\rceil$. Apart from this, formulae are built up in the usual way. Formulae not containing free variables are called "sentences".

Metalanguage: We denote formulae by Gothic capitals, and use the following convention to indicate parts of a formula: $\mathfrak{A}[A(a), B(b)]$ says that $\mathfrak{A}$ contains at most the indicated prime formulae (and perhaps the signs T,F) and is built up by means of sentential connectives only; in opposite, $\mathfrak{A}(A(a),B(b))$ may be any formula containing at least the indicated parts. For example, the formula $(\exists t)\lceil A(t) \wedge \lceil B_1(a) \vee B_2(a)\rceil\rceil$ may be shortened as $(\exists t)\mathfrak{A}[A(t),\underline{B}^2(a)]$, $(\exists t)\mathfrak{A}[A(t),\underline{B}(a),D(e)]$, $\mathfrak{A}(a)$, $\mathfrak{A}(\underline{B})$ or $(\exists t)\mathfrak{A}(t)$. By $\equiv$ we indicate semiotic equality of formulae. - To denote natural numbers (needed as indices for variables, and to indicate the length of conjunctions, the number of variables, etc.) we use the same signs $\mathfrak{i},...,\mathfrak{v}$ as above. Sets and relations of natural numbers will be indicated by the corresponding Latin capitals: I,J,...,N. As for the rest we use the signs of the object language in the metalanguage. Further conventions will be introduced when needed.

Logical axioms: We use freely axioms and rules of propositional calculus without mentioning. We get the other rules and axioms by simply writing down, firstly an axiom system of the ordinary predicate calculus, and secondly a copy of which the individual variables are replaced by predicate variables (cf. Hilbert-Bernays [20], II, p.500f). Thus we state the other axioms and rules by pairs (respectively for individual and predicate variables). It is to be understood that one has to avoid collision of variables.

1) Substitution rule:

(SI) $\dfrac{\mathfrak{A}(a)}{\mathfrak{A}(\alpha)}$

(where α is a term)

(SP) $\dfrac{\mathfrak{A}(A)}{\mathfrak{A}(\mathfrak{L})}$

(where $\mathfrak{L}$ is a formula with one marked free individual variable)

2) Changing of bound variables:

(CI) $\dfrac{(\Delta x)\,\mathfrak{A}(x)}{(\Delta y)\,\mathfrak{A}(y)}$

(where Δ is a quantifier)

(CP) $\dfrac{(\Delta P)\,\mathfrak{A}(P)}{(\Delta R)\,\mathfrak{A}(R)}$

3) Axioms for quantifiers:

(AQI1) $(\forall x)\,\mathfrak{A}(x) \to \mathfrak{A}(a)$

(AQI2) $\mathfrak{A}(a) \to (\exists x)\,\mathfrak{A}(x)$

(AQP1) $(\forall P)\,\mathfrak{A}(P) \to \mathfrak{A}(A)$

(AQP2) $\mathfrak{A}(A) \to (\exists P)\,\mathfrak{A}(P)$

4) Rules for quantifiers:

(RQI1) $\dfrac{\mathfrak{L} \to \mathfrak{A}(a)}{\mathfrak{L} \to (\forall x)\,\mathfrak{A}(x)}$

(RQP1) $\dfrac{\mathfrak{L} \to \mathfrak{A}(A)}{\mathfrak{L} \to (\forall P)\,\mathfrak{A}(P)}$

(RQI2) $\dfrac{\mathfrak{A}(a) \to \mathfrak{b}}{(\exists x)\,\mathfrak{A}(x) \to \mathfrak{b}}$ (RQP2) $\dfrac{\mathfrak{A}(A) \to \mathfrak{b}}{(\exists P)\,\mathfrak{A}(P) \to \mathfrak{b}}$

(a not in $\mathfrak{b}$)   (A not in $\mathfrak{b}$)

Evidently the axiom system is not independent.

We call this logical frame $P^1K(2)$: second order predicate calculus with one-place predicate variables, or shorter monadic second order predicate calculus. A theory formalized within this logic will be called a monadic second order theory.

It is wellknown that within this frame equality is definable by

$$a = b \iff_{df} (\forall P)\big\lceil P(a) \to P(b)\big\rceil$$

Further a special form of the replacement theorem is derivable, which we call principle of extensionality:

(EXT) $(\forall x)\big\lceil A(x) \iff B(x)\big\rceil \,.\!\to\!.\; \mathfrak{A}(A) \iff \mathfrak{A}(B)\,.$

The derivation uses induction over the length of the formula $\mathfrak{A}$, and does not involve (SP). By the same method, or using the ordinary replacement theorem of the propositional calculus, we get the generalization

(1) $(\forall x)\big\lceil A(x) \iff \mathfrak{b}(x)\big\rceil \,.\!\to\!.\; \mathfrak{A}(A) \iff \mathfrak{A}(\mathfrak{b})\,.$

With the help of this formula we show that the substitution rule (SP) is equivalent to the principle of comprehension (see Henkin [18]):

(COMP) $(\exists P)(\forall x)\big\lceil P(x) \iff \mathfrak{A}(x)\big\rceil$ (P not in $\mathfrak{A}$),

in the sense that each of them is derivable from the other. In view of the later fundamental rôle of (COMP) let us present this proof: On one hand, from

$(\forall x)\big\lceil A(x) \iff \mathfrak{A}(x)\big\rceil \to (\exists P)(\forall x)\big\lceil P(x) \iff \mathfrak{A}(x)\big\rceil$,

an instance of (AQP2), we get by (SP)

$(\forall x)\big\lceil \mathfrak{A}(x) \iff \mathfrak{A}(x)\big\rceil \to (\exists P)(\forall x)\big\lceil P(x) \iff \mathfrak{A}(x)\big\rceil$

which yields directly (COMP). On the other hand, from (1) we derive

$(\exists P)(\forall x)\big\lceil P(x) \iff \mathfrak{b}(x)\big\rceil \,.\!\to\!.\; (\exists P)\big\lceil \mathfrak{A}(P) \iff \mathfrak{A}(\mathfrak{b})\big\rceil$,

using (AQP2) and (RQP2). By (COMP) we get rid of the antecedent of this conditional; from the consequent follows easily

$$(\forall P)\,\mathfrak{A}(P) \to \mathfrak{A}(\mathfrak{b})\,.$$

By (RQP1), this yields $\mathfrak{A}(\mathfrak{b})$ from the premise $\mathfrak{A}(A)$; thus (SP) is derived. - By the presence of (COMP), $P^1K(2)$ may be considered as a fragment of set theory (cf. R.M.Robinson[33], Hasenjaeger[17], Mc Naughton [22]). In most derivations we will use (COMP) instead of SP, and it is in fact this highly impredicative comprehension principle which gives together with the induction axiom the strength of SC.

<u>Non-logical axioms:</u> The three Peano axioms for successor are sufficient.
We need no schema in view of (SP):

(A1) $a' = b' \to a = b$

(A2) $a' \neq o$

(I) $A(o) \wedge (\forall t)\lceil A(t) \to A(t')\rceil \to (\forall t)A(t)$.

The theory built up in $P^1K(2)$ with these non-logical signs and by these
non-logical axioms, is called <u>sequential calculus SC</u>.

It will be our task in this chapter to show that SC is decidable, i.e.
that the set of Gödel numbers of the derivable sentences of SC is recursive (see e.g. Hermes[19]). The common method is to present a mechanical
procedure which transforms any given SC-sentence into T or F according
to whether it is true or false; in our case this procedure involves only
derivable equivalences. Therefore we explore in §1 the power of SC.

First of all we get from (I) by (SP) the induction schema

(IS) $\mathcal{O}(o) \wedge (\forall t)\lceil \mathcal{O}(t) \to \mathcal{O}(t')\rceil \to (\forall t)\mathcal{O}(t)$.

Further it is known (cf.Hilbert-Bernays[20],II,p.501f; for illustration
see the remarks on switching predicates below) that order is definable
by

$a < b \iff_{df} (\exists P).P(a) \wedge (\forall t)\lceil P(t') \to P(t)\rceil \wedge \neg P(b)$.

We use the following abbreviations (cf.Büchi[3]):

1) $(\exists t)_a^b \mathcal{O}(t) \iff_{df} (\exists t)\lceil a \leq t < b \wedge \mathcal{O}(t)\rceil$

2) $(\forall t)_a^b \mathcal{O}(t) \iff_{df} (\forall t)\lceil a \leq t < b \to \mathcal{O}(t)\rceil$

3) $(\exists^{\smile} t) \mathcal{O}(t) \iff_{df} (\forall x)(\exists t)\lceil x < t \wedge \mathcal{O}(t)\rceil$

4) $(\forall^{\smile} t) \mathcal{O}(t) \iff_{df} (\exists x)(\forall t)\lceil x < t \to \mathcal{O}(t)\rceil$

5) $(\exists P)^{\smile} \mathcal{O}(P) \iff_{df} (\exists P)\lceil (\exists^{\smile} t)P(t) \wedge \mathcal{O}(P)\rceil$.

1) and 2) are the familiar restrictions of quantifiers; sometimes we
use also $(\exists t)_a \mathcal{O}(t)$ and $(\forall t)_a \mathcal{O}(t)$, if there is only a lower bound. 3)
is to be read as "there are infinitely many t such that $\mathcal{O}(t)$", 4) as
"for ultimately all t, $\mathcal{O}(t)$", 5) as "there is an infinite P such that
$\mathcal{O}(P)$". Remark that the following formulae are derivable:

1) $\neg(\exists t)_a^b \mathcal{O}(t) \iff (\forall t)_a^b \neg \mathcal{O}(t)$

2) $(\forall^{\smile} t) \mathcal{O}(t) \to (\exists^{\smile} t) \mathcal{O}(t)$

3) $\neg(\exists^{\smile} t) \mathcal{O}(t) \iff (\forall^{\smile} t) \neg \mathcal{O}(t)$.

At last we define for any $\mathcal{M} > o$

$$a \equiv b \ (\textit{\textrm{n}}) \ <->_{df} \ (\forall P).P(a) \ \wedge \ (\forall t)\lceil P(t) \ <-> \ P(t+\textit{\textrm{n}})\rceil \ -> \ P(b)$$

(" a is congruent b modulo *n*").

Evidently, SC is a fragment of numbertheory. Thus, if $\mathbb{N}$ is the set of natural numbers - given anyhow -, $\mathbb{N}$ becomes a model of SC under the usual interpretation. E.g. predicate variables are interpreted as functions from $\mathbb{N}$ into the set $\{T,F\}$, i.e. as ω-sequences of T's and F's; we call such sequences predicates. - The word "system" may denote an interpreted theory. Thus SC under the above interpretation is a <u>monadic second order system</u> of arithmetic.

It is very convenient for understanding the meaning of SC-formulae to think of the elements of $\mathbb{N}$ as of points in a discrete time scale. Then a predicate is fancied best as representing a thing (e.g. a light) which can vary in time between two values T,F ("on","off"). The course in time of such a thing and of finite systems of such things may be called a "process". A lot of events in time can be brought into the form of such processes, examples are: (1) the varying look of the windows (whether lightened or not) in the skyline of New York, (2) the behaving of a computer (where the two-valued things are the transistors, lights and so on), (3) the working of a nerve net. Example (1) suggests to represent processes by ω-sequences of o-1-matrices (as is often done, and as we will do later on, see 2.a). Example (2) explains the origin of the name "SC", "sequential calculus": Büchi set up his system SC to examine certain electric circuits, called sequential circuits, systems of lights and switches which change their state synchronously in discrete time. But the very predecessors of SC are the neuro-biologists (example 3), who used firstly processes (but not the name, as it seems) as models of the behaving of nerve nets. For details see II.1.a. For the moment we note only that for our purposes a process is a <u>finite</u> system of things each of which is capable of two states (which is clearly no restriction against the general case of finitely many states); the things change their states synchronously in discrete time steps, the states together determine the state of the system; the life of the system is thought to begin at a certain moment with a certain state, but it is <u>infinite</u> in time. To say it mathematically: a process is a finite set of functions from $\mathbb{N}$ into the set $\{T,F\}$; thus a process may be represented by a tuple of predicates.

Examples (3) and (2) show also the most useful application of SC: Commonly, a process is not free, but is steered by certain informations from the outside, or in other words it is governed by certain conditions. So one may ask what processes result from given conditions, or especially whether there are processes at all satisfying the conditions. The investigation of this point has led to the theory of finite automata, which is touched in II.1.a where also references are given. Here the point of interest for us is that, under this interpretation, SC-formulae containing free predicate variables become conditions on processes. For example, the formula $(\forall t)A(t)$ determines the sequence consisting of T's only. The formula $(\forall t)\lceil A(t) <-> A(t')\rceil$ determines a "stable" predicate which never changes its state, thus is always T or always F. We shall use the idea of stable predicates sometimes in the following to link two different processes (see the proofs of theorem d.1 and 3.b.1). Similarly, we shall use often the idea of a "switching predicate", i.e. a predicate defined to be false up to a certain point, "switched on" at this point (e.g. by a certain event) and staying to be true – or the same with T and F interchanged. Switching predicates are defined for example by the formulae $(\forall t)\lceil A(t') \rightarrow A(t)\rceil$ and $(\forall t)\lceil A(t) \rightarrow A(t')\rceil$, if we regard stable predicates as degenerated switching predicates. An example of this idea is the above definition of the smaller relation: a < b if and only if there exists a switching predicate turned off between a and b . Thus the numbertheoretic weakness of SC is compensated to a certain amount by the power of the second order logic: we have nothing but the successor function to get along number by number, but we have predicates to collect and to transmit information with the help of the successor function.

To illustrate the usefullness of this point of view we collect a number of equivalences derivable in SC, which are understood best by the idea of switching predicates. Namely, if $\mathcal{O}(a)$ is any formula not containing the predicate variable A, then the formula

$$\neg A(o) \wedge (\forall t)\lceil A(t') <-> A(t) \vee \mathcal{O}(t)\rceil$$

is easily seen to determine a predicate which is F up to the smallest number n so that $\mathcal{O}(n)$ holds, and T above n (or F everywhere if such an n does not exist). We say that A is "switched on by $\mathcal{O}$ at n".

Remarks:

1) $(\exists x)A(x) :<->: (\forall P).\neg P(o) \wedge (\forall t)\lceil P(t') <-> P(t) \vee A(t)\rceil \rightarrow (\exists t)P(t)$

2) $(\exists x)A(x) :<->: (\exists P).\neg P(o) \wedge (\forall t)\lceil P(t') <-> P(t) \vee A(t)\rceil \wedge (\exists t)P(t)$

3) $(\exists x)A(x) :<->: (\exists P).\neg P(o) \wedge (\forall t)\lceil P(t') <-> P(t) \vee A(t)\rceil \wedge (\exists^w t)P(t)$

4) $(\exists x)A(x) :<->: (\exists P).\neg P(o) \wedge (\forall t)\lceil P(t') <-> P(t) \vee A(t)\rceil \wedge (\forall^w t)P(t)$

5) $\bigwedge_{i=1}^{m} (\exists x)A_i(x) :<->: (\exists \underline{P}^m). \bigwedge_{i=1}^{m} \neg P_i(o) \wedge (\forall t)\lceil \underline{P}(t') <-> \underline{\mathcal{G}}(t)\rceil \wedge (\exists^w t) \bigwedge_{i=1}^{m} P_i(t)$

   (where $\mathcal{G}_i(a) \equiv_{df} P_i(a) \vee A_i(a), i = 1,\ldots,m$)

6) $a \leq b .:->:. (\exists x)_a^b A(x) :<->: (\exists P).\neg P(a) \wedge (\forall t)_a^b \lceil P(t') <-> P(t) \vee A(t)\rceil \wedge P(b)$

7) $(\exists t)A(t) \wedge (\exists^w t)B(t) :<->: (\exists P).\neg P(o) \wedge$

$\wedge (\forall t)\lceil P(t') <-> P(t) \vee A(t)\rceil \wedge (\exists^w t)\lceil P(t) \wedge B(t)\rceil$

Here we have widened in remark 5 our convention on writing n-tuples (see beginning of the section) to recursion equivalences; we shall use recursion so often that an abbreviation will be justified. Only for these types of formulae we will denote n-tuples of formulae by under-lining, and will write

$$\underline{B}^n(a') <-> \underline{\mathfrak{L}}^n(a)$$

instead of

$$\bigwedge_{i=1}^{n} \lceil B_i(a') <-> \mathfrak{L}_i(a)\rceil \quad .$$

We get the proof of remark 1 by induction. Remarks 5) - 7), which will be particularly useful in the sequel, follow easily from remarks 2) - 4), the right sides of which are easily shown to be pairwise equivalent. But to prove e.g. remark 2) we have to derive in SC the formula

$$(\exists P).\neg P(o) \wedge (\forall t)\lceil P(t') <-> P(t) \vee A(t)\rceil \quad .$$

We will show in the next section the more general theorem that the existence of predicates determined by simultaneous course-of-value re-cursion is derivable in SC. Having done this we have at hand a powerful instrument to get proofs in SC, and indeed to show the completeness of our axiom system. Namely in section c we will derive with the help of the recursion theorem the theorem of Ramsey which is used in Büchi's DP.

These examples give an first account of how SC-formulae determine processes. E.g. the equivalence of remark 5 has the meaning: A process has in each element a positive point iff there is another process con-sisting of switching predicates, each of which begins "out", and is switched on by an element of the first process, and which all have infi-nitely many (and indeed ultimately all) points in common where they are "on". Within this section we will use this manner of speaking only to illustrate formal proofs. Later on we will ask for the processes deter-mined by types of formulae. It will be the deepest consequence of the decidability of SC that processes determined by SC-formulae are ulti-mately periodic, and thus are , though infinite, not better than finite processes. More precisely: If a condition expressed in the language of SC is satisfied by a process at all, then there is also a process satis-fying this condition which up from a certain point repeats at infinity a certain finite part.

We conclude with a remark on the presentation of proofs: We write down proofs in the usual half-formal way, e.g. using free variables to denote its own interpretation, sc. numbers and predicates. In most cases, we hope, it should be easy to get strictly formal proofs from the given ones. In §5, however, we shall give a more formal version of some difficult proofs. It turns out there that some of these informal considerations are not as innocent as they look.

b) The recursion theorem

We have maintained in section a that the existence of predicates introduced by recursion equivalences is derivable in SC. We will show now this "recursion theorem" in a slightly more general way, viz. for course-of-value recursion. For the sake of simplicity we will give the proof for one predicate only, indicating afterwards how to generalize to simultaneous recursion.

For the whole section, let $\mathcal{L}(a, E(t); t < a)$ mean that in the formula $\mathcal{L}$ the predicate variable E is contained only with bound arguments and that all these arguments are restricted by a . $\mathcal{L}$ may contain other free and bound variables. For such a formula $\mathcal{L}(a, E)$ we have a restricted form of (EXT):

(REXT) $(\forall x)_o^a \lceil A(x) \iff B(x) \rceil \ .\to. \ \mathcal{L}(a, A) \iff \mathcal{L}(a, B)$.

We want to show that the course-of-value recursion

(R) $E(a) \iff \mathcal{L}(a, E(t); t < a)$

defines a predicate in SC, that means that the existence of such a predicate is derivable. Clearly this cannot be done by (COMP) alone, since E is contained in $\mathcal{L}$, too. So we have to do a little more. - Of course, the proof of this "recursion theorem" follows the known pattern of the proof of the recursion theorem e.g. in set theory.

Note that, by our definition of $\mathcal{L}(a, E)$, $\mathcal{L}(o, E(t); t < o)$ is equivalent to a formula $\lceil$ not containing E at all. Thus we have by (COMP) $(\exists P)(\forall t) \lceil P(t) \iff \lceil \rceil$ and therefore $(\exists P) \lceil P(o) \iff \lceil \rceil$, which ensures us the existence of a predicate starting as we want it. (Of course we could have stated the whole problem with an extra initial condition.)

First we show the uniqueness of any initial part of a recursively defined predicate:

__Lemma 1:__ $(\forall z)_o^b \lceil E(z) \iff \mathcal{L}(z, E) \rceil \wedge (\forall z)_o^c \lceil D(z) \iff \mathcal{L}(z, D) \rceil \wedge b \leq c \to$
$$\to (\forall z)_o^b \lceil E(z) \iff D(z) \rceil \quad .$$

__Proof:__ By induction (IS) over b : The beginning b = o is trivial. From the premise for b' follows the premise for b , and thus by induction hypothesis the conclusion for b : $(\forall z)_o^b \lceil E(z) \iff D(z) \rceil$. From this we get by (REXT) above $\mathcal{L}(b, E) \iff \mathcal{L}(b, D)$ and therefore $E(b) \iff D(b)$, which gives the conclusion for b'. #

Next we derive the existence of any initial part of a recursively defined predicate:

__Lemma 2:__ $(\exists S)(\forall z)_o^{b'} \lceil S(z) \iff \mathcal{L}(z, S) \rceil$.

Proof: Induction over b : The zero case is trivial (cf. the above remark).

Induction step: By induction hypothesis we have a predicate B so that

$$(\forall z)_{o}^{a'} \big\lceil B(z) \Longleftrightarrow \mathfrak{L}(z,B) \big\rceil \quad .$$

We abbreviate this formula by $\mathfrak{h}^{\mathfrak{L}}(a,B)$. By (COMP) we get

$$(\exists S)(\forall z).S(z) \Longleftrightarrow \big\lceil z \leq b \wedge B(z) \big\rceil \ \vee \ \big\lceil z = b' \wedge \mathfrak{L}(b',B) \big\rceil$$

and therefore

(1) $(\exists S).(\forall z)_{o}^{b'} \big\lceil S(z) \Longleftrightarrow B(z) \big\rceil \ \wedge \ \big\lceil S(b') \Longleftrightarrow \mathfrak{L}(b',B) \big\rceil \quad .$

From the formula

$$(\forall z)_{o}^{b'} \big\lceil D(z) \Longleftrightarrow B(z) \big\rceil$$

follows by (REXT)

$$\big\lceil \mathfrak{h}^{\mathfrak{L}}(b,B) \Longleftrightarrow \mathfrak{h}^{\mathfrak{L}}(b,D) \big\rceil \ \wedge \ \big\lceil \mathfrak{L}(b',B) \Longleftrightarrow \mathfrak{L}(b',D) \big\rceil \quad .$$

Thus we have by propositional calculus

$$\mathfrak{h}^{\mathfrak{L}}(b,B) \wedge (\forall z)_{o}^{b'} \big\lceil D(z) \Longleftrightarrow B(z) \big\rceil \wedge \big\lceil D(b') \Longleftrightarrow \mathfrak{L}(b',B) \big\rceil \rightarrow$$
$$\rightarrow \mathfrak{h}^{\mathfrak{L}}(b,D) \wedge \big\lceil D(b') \Longleftrightarrow \mathfrak{L}(b',D) \big\rceil \quad .$$

By quantification we get

$$\mathfrak{h}^{\mathfrak{L}}(b,B) \wedge (\exists S) \Big\lceil (\forall z)_{o}^{b'} \big\lceil S(z) \Longleftrightarrow B(z) \big\rceil \wedge \big\lceil S(b') \Longleftrightarrow \mathfrak{L}(b',B) \big\rceil \Big\rceil \rightarrow$$
$$\rightarrow (\exists S) \, \mathfrak{h}^{\mathfrak{L}}(b',S) \quad .$$

From this follows by (1)

$$\mathfrak{h}^{\mathfrak{L}}(b,B) \rightarrow (\exists S) \, \mathfrak{h}^{\mathfrak{L}}(b',S) \quad ,$$

which leads to the wanted conclusion

$$(\exists S) \, \mathfrak{h}^{\mathfrak{L}}(b,S) \rightarrow (\exists S) \, \mathfrak{h}^{\mathfrak{L}}(b',S) \quad . \quad \#$$

Now we are able to prove our quoted theorem:

Theorem 1 (recursion theorem): For any formula $\mathfrak{L}(a,E(t); \ t < a)$ the following is derivable in SC:

$$(\exists s)(\forall z) \big\lceil S(z) \Longleftrightarrow \mathfrak{L}(z,S) \big\rceil \quad .$$

Proof: Let $\mathfrak{h}^{\mathfrak{L}}$ be defined as in the proof of lemma 2. By (COMP) we have

$$(\exists S)(\forall z).S(z) \Longleftrightarrow (\exists R) \big\lceil \mathfrak{h}^{\mathfrak{L}}(z,R) \wedge R(z) \big\rceil \quad .$$

Let us abbreviate this formula by

(1) $(\exists S) \, \vartheta(S) \quad .$

It is easy to see that with the help of lemma 1 we have

$$\mathfrak{h}^{\mathfrak{L}}(b,C) \wedge \vartheta(D) \ . \rightarrow . \ (\forall t)_{o}^{b'} \big\lceil C(t) \Longleftrightarrow D(t) \big\rceil \quad ,$$

especially

$$\mathcal{h}^{\mathcal{b}}(b,C) \wedge \vartheta(D) \; .\to. \; C(b) \iff D(b) \; .$$

By (REXT) we have

$$(\forall t)_o^{b'} \big[C(t) \iff D(t) \big] \; .\to. \; \mathcal{L}(b,C) \iff \mathcal{L}(b,D)$$

and by definition of $\mathcal{h}^{\mathcal{b}}$

$$\mathcal{h}^{\mathcal{b}}(b,C) \; .\to. \; C(b) \iff \mathcal{L}(b,C) \; .$$

Together we get

$$\mathcal{h}^{\mathcal{b}}(b,C) \wedge \vartheta(D) \; .\to. \; D(b) \iff \mathcal{L}(b,D) \; .$$

Quantification gives us step by step

$$(\exists R)\, \mathcal{h}^{\mathcal{b}}(b,R) \wedge \vartheta(D) \; .\to. \; D(b) \iff \mathcal{L}(b,D) \quad ,$$

$$(\forall z)(\exists R)\, \mathcal{h}^{\mathcal{b}}(z,R) \wedge \vartheta(D) \; .\to. \; (\forall z)\big[D(z) \iff \mathcal{L}(z,D) \big] \quad ,$$

$$(\forall z)(\exists R)\, \mathcal{h}^{\mathcal{b}}(z,R) \wedge (\exists S)\vartheta(S) \; .\to. \; (\exists S)(\forall z)\big[S(z) \iff \mathcal{L}(z,S) \big] \quad .$$

With the help of lemma 2 and (1) we have our theorem. $\#$

Three remarks to conclude: 1) It is clear that the recursion theorem holds for ordinary recursion, say

(1)
$$E(o) \iff T$$
$$E(a') \iff \mathcal{L}(E(a),\underline{A}^{\mathcal{n}}(a))$$

For (1) can be transformed into the form (R), e.g.

$$E(a) \iff a = o \vee (\exists z)_o^a \big[z' = a \wedge \mathcal{L}(E(z),\underline{A}(z)) \big].$$

If we have F instead of T in the first equivalence, we drop the disjunct a = o.

2) The recursion theorem holds as well for simultaneous recursion of two or more predicates

(2) $$\underline{E}^{\mathcal{n}}(a) \iff \underline{\mathcal{L}}^{\mathcal{n}}(a,\underline{E}(t);\ t < a)$$

(For notation see section a.) The proof carries over nearly unchanged from the ordinary case. Just as in ordinary recursion theory, the form (2) can again be generalized. For example, in the next section we will use the recursion theorem for a formula of the following type

(3)
$$D(a) \iff \vartheta(a,D(t),E(t);\ t < a)$$
$$E(a) \iff \mathcal{f}(a,D(a),D(t),E(t);\ t < a)$$

3) For latter application we note that the uniqueness of predicates introduced recursively by any of these forms is easily derived in SC – just as it was done in lemma 1 for the initial parts of such predicates.

c) The theorem of Ramsey

As a first example of the usefulness of the recursion theorem, in this section we will derive in SC as theorem 3 the formal counterpart of Ramsey's theorem (reformulated in theorem 1 below). This shows at once the relative power of SC. Moreover, in section 2.d we will apply the theorem of Ramsey to certain special formulae (the so-called Σ°-formulae defined in 2.a), thus solving the main problem in getting a decidable normal form for SC-sentences. In doing so we get another formulation of the Ramsey theorem, which originates from the application of this theorem in the paper [3] of Büchi.

For any natural number $n > o$ and any set M let $\pi_n(M)$ be the set of all n-element subsets of M. Then theorem A of Ramsey [31] reads as follows:

Theorem 1 (Ramsey): Let N be an infinite set, let n, m be natural numbers, let $L_1, \ldots, L_m$ be a partition of $\pi_n(N)$ (i.e. the L_i are disjunct, and their union is $\pi_n(N)$). Then there exists a number f, $1 \leq f \leq m$, and an infinite subset M of N so that $\pi_n(M) \subseteq L_f$.

To prove this theorem in SC means first that we choose as N the set $\mathbb{N}$ of natural numbers. For the beginning we will further restrict us to the case $n = 2$ and $m = 2$ (we will treat the general case at the end of the section). Since we have a wellordered domain, we may speak of sequences instead of subsets and of "ascending" pairs (a,b) (i.e. a < b) instead of two-element subsets $\{a,b\}$. Then the theorem of Ramsey says: For any partition L_1, L_2 of the ascending pairs of natural numbers there exists an infinite sequence so that every ascending pair chosen from the sequence lies in one and the same partition set.

It is possible to formalize the original proof of Ramsey within SC; this is done in the paper [36]. Later on we learned from Dr.R.B. Jensen (Bonn) another proof of the Ramsey theorem which is shorter than the usual one. Moreover this proof is much easier to translate into the language of SC, and its derivation is nearly trivial by means of the recursion theorem. So we give this proof with Dr. Jensen's kind permission. (See also Kuratowski-Mostowski, "Set Theory", pp.1o7-1o9.)

We begin with an informal version of the proof, referring at each step to the subsequent formalization. - So let L_1, L_2 be a partition of the ascending pairs of natural numbers. We want to show that for $i = 1$ or $i = 2$ there must exist an infinite L_i-sequence, i.e. an infinite sequence every ascending pair of which lies in L_i. For the proof we use the following trivial fact: Let M be an infinite set of numbers, let a be a fixed number. Define $M_i =_{df} \{x \varepsilon M; (a,x) \varepsilon L_i\}$, $i = 1,2$. Then M_1 or

M_2 is infinite.

Now we define recursively an infinite set $\{a_0, i(0), a_1, i(1), a_2, i(2),$ $...\}$ where each $i(j)$ is 1 or 2. More exactly, we define recursively the domain $\{a_0, a_1, ...\}$ and the values (from $\{1,2\}$) of a "label"-function g in the following way:

$$g(o) \ =_{df} \ \begin{cases} 1; \ (\exists^\omega t) \ \lceil o < t \wedge (o,t)\varepsilon L_1 \rceil \\ \\ 2; \ \text{otherwise} \end{cases}$$

$$a_1 \ =_{df} \ \min \ \{x; \ o < x \wedge (o,x)\varepsilon L_{g(o)} \}$$

$$g(a_1) \ =_{df} \ \begin{cases} 1; \ (\exists^\omega t) \ \lceil a_1 < t \wedge (a_1,t)\varepsilon L_1 \wedge (o,t)\varepsilon L_{g(o)} \rceil \\ \\ 2; \ \text{otherwise} \end{cases}$$

And so on. In general, let g be defined for $a_0 =_{df} o, a_1, ..., a_\ell$:

$$a_{\ell+1} \ =_{df} \ \min \{x; \ a_\ell < x \wedge \bigwedge_{i=o}^{\ell} \ (a_i,x)\varepsilon L_{g(a_i)} \}$$

$$g(a_{\ell+1}) \ =_{df} \ \begin{cases} 1; \ (\exists^\omega t) \ \lceil a_{\ell+1} < t \wedge (a_{\ell+1},t)\varepsilon L_1 \wedge \bigwedge_{i=o}^{\ell} \ (a_i,t)\varepsilon L_{g(a_i)} \rceil \\ \\ 2; \ \text{otherwise} \end{cases}$$

Clearly, this definition makes sense only if one can show that for any ℓ there exists a number a so that $a_\ell < a \wedge \bigwedge_{i=o}^{\ell} (a_i,a)\varepsilon L_{g(a_i)}$. This is done in the sequel by lemma 2. The proof of lemma 2 is trivial if one uses the fact that even

$$(\exists^\omega t).a_\ell < t \wedge \bigwedge_{i=0}^{\ell} (a_i,t)\varepsilon L_{g(a_i)}$$

is true for any ℓ. Namely this is clear for $\ell = o$, and if the said formula is shown for ℓ, the remark of the preceding paragraph yields directly the formula for $\ell+1$. The formalization of this proof is given in lemma 1 below. - From the definition of g follows immediately that $(a_i,a_j)\varepsilon L_{g(a_i)}$ for any pair $i < j$. Especially, any ascending pair from $g^{-1}\{i\}$ lies in L_i, $i = 1,2$ (lemma 3). Since by construction the domain of g is infinite, either $g^{-1}\{1\}$ or $g^{-1}\{2\}$ is infinite, and satisfies the conclusion of the theorem of Ramsey. - Clearly this way of defining a "label"-function is well-known, e.g. from the construction of non-standard models by Skolem and from the Gödel completeness theorem.

To formulate theorem 1 in the language of SC we have to express sets by predicates. Since SC has only one-place predicate variables, we cannot write down the theorem for arbitrary, but only for definable partition sets. So let $\mathcal{F}(a,b)$ be an SC-formula which may contain other individual- and predicate variables. Then we translate the special case of the Ramsey theorem into the language of SC by the following formula (it is just this formula for Σ^o-formulae $\mathcal{F}(a,b)$ Büchi suggested as axiom schema for SC):

Theorem 2: $(\exists Q)^\omega \cdot (\forall y)(\forall x)_o^y \lceil Q(x) \wedge Q(y) \rightarrow \mathcal{h}(x,y) \rceil$ v

$\qquad\qquad$ v $(\forall y)(\forall x)_o^y \lceil Q(x) \wedge Q(y) \rightarrow \neg \, \mathcal{h}(x,y) \rceil$.

For the formalization of the above proof we have to replace $(a,b) \epsilon L_1$ by $\mathcal{h}(a,b)$ and $(a,b) \epsilon L_2$ by $\neg \, \mathcal{h}(a,b)$. Further we use predicate variables A and B to represent the domain of g and the set $g^{-1}\{1\}$, respectively. For example, the part $(a_i, x) \epsilon L_{g(a_i)}$ in the definition of a_{i+1} becomes $A(a_i) \rightarrow \lceil B(a_i) <-> \mathcal{h}(a_i,x) \rceil$. So the following formulae $\mathcal{L}$ and $\mathcal{L}$ are direct translations of the definitions of the a_i and of g :

$$\mathcal{L}(A,B) \equiv_{df} (\forall x) \lceil A(x) <-> (\forall z)_o^x \lceil A(z) \rightarrow \lceil B(z) <-> \mathcal{h}(z,x) \rceil \rceil \rceil ,$$

$$\mathcal{L}(A,B) \equiv_{df} (\forall x) \lceil B(x) <-> A(x) \wedge (\exists^\omega t) \lceil x < t \wedge \mathcal{h}(x,t) \wedge$$
$$\wedge (\forall z)_o^x \lceil A(z) \rightarrow \lceil B(z) <-> \mathcal{h}(z,t) \rceil \rceil \rceil \rceil .$$

To get lemma 2 we define a third formula

$$\vartheta(A,B,a,b) \equiv_{df} a < b \wedge (\forall z)_o^{a'} \lceil A(z) \rightarrow \lceil B(z) <-> \mathcal{h}(z,b) \rceil \rceil ,$$

and show:

Lemma 1: $\mathcal{L}(A,B) \wedge \mathcal{L}(A,B) \rightarrow (\forall z) \lceil A(z) \rightarrow (\exists^\omega t) \vartheta(A,B,z,t) \rceil$.

Proof: Let A,B satisfy $\mathcal{L}(A,B) \wedge \mathcal{L}(A,B)$. We shall show by induction on a:

$$A(a) \rightarrow (\exists^\omega t) \vartheta(A,B,a,t) .$$

Beginning: a = o. Since A(o) holds, we have to prove:

$$(\exists^\omega t) . o < t \wedge \lceil B(o) <-> \mathcal{h}(o,t) \rceil .$$

This follows from

$$B(o) <-> (\exists^\omega t) \lceil o < t \wedge \mathcal{h}(o,t) \rceil ,$$

which holds by the premise $\mathcal{L}(A,B)$.

For the induction step let a be a number so that $o < a \wedge A(a)$. Let b be the greatest number so that $b < a \wedge A(b)$; by induction hypothesis we have

$$(\exists^\omega t) \vartheta(A,B,b,t) .$$

1. case: B(a). From $\mathcal{L}(A,B)$ follows

$$(\exists^\omega t) . a < t \wedge \mathcal{h}(a,t) \wedge (\forall z)_o^a \lceil A(z) \rightarrow \lceil B(z) <-> \mathcal{h}(z,t) \rceil \rceil ,$$

which gives together with B(a)

$$(\exists^\omega t) . a < t \wedge (\forall z)_o^{a'} \lceil A(z) \rightarrow \lceil B(z) <-> \mathcal{h}(z,t) \rceil \rceil .$$

2. case: $\neg B(a)$. In view of $(\forall t)_b^a \neg A(t)$ we infer from $(\exists^\omega t) \vartheta(A,B,b,t)$

(1) $(\exists^\omega t) . a < t \wedge (\forall z)_o^a \lceil A(z) \rightarrow \lceil B(z) <-> \mathcal{h}(z,t) \rceil \rceil .$

From ¬B(a) we get further by $\mathcal{L}(A,B)$

(2) $(\forall^\omega t). a < t \wedge (\forall z)_o^a \lceil A(z) \rightarrow \lceil B(z) <\rightarrow \mathcal{G}(z,t) \rceil \rceil \rightarrow \neg \mathcal{G}(a,t)$.

(1) and (2) together yield

$(\exists^\omega t). a < t \wedge \neg \mathcal{G}(a,t) \wedge (\forall z)_o^a \lceil A(z) \rightarrow \lceil B(z) <\rightarrow \mathcal{G}(z,t) \rceil \rceil$,

which gives $(\exists^\omega t) \vartheta(A,B,a,t)$. #

Lemma 2: $\mathcal{L}(A,B) \wedge \mathcal{L}(A,B) \rightarrow (\exists^\omega t)A(t)$

Proof: Let A,B satisfy $\mathcal{L}(A,B) \wedge \mathcal{L}(A,B)$. Since A(o) holds, it is
sufficient to show:

$(\forall x).A(x) \rightarrow (\exists y)\lceil x < y \wedge A(y) \rceil$.

So let a be a number so that A(a).
From lemma 1 follows

$(\exists^\omega t) \vartheta(A,B,a,t)$

Let b be the smallest number so that $\vartheta(A,B,a,b)$. We want to show:

$(\exists t)_{a'}^{b'}A(t)$.

If $(\exists t)_a^b A(t)$ holds, we are ready. If on the other hand $(\forall t)_{a'}^b \neg A(t)$ is
true, from

$(\forall z)_o^{a'} \lceil A(z) \rightarrow \lceil B(z) <\rightarrow \mathcal{G}(z,b) \rceil \rceil$

— which is a part of $\vartheta(A,B,a,b)$ — we get

$(\forall z)_o^b \lceil A(z) \rightarrow \lceil B(z) <\rightarrow \mathcal{G}(z,b) \rceil \rceil$,

i.e. A(b). #

    From the definition of $\mathcal{L}$ and $\mathcal{L}$ follows immediately that the so
fixed predicates A and B do what we want:

Lemma 3: $\mathcal{L}(A,b) \wedge \mathcal{L}(A,B) \wedge A(a) \wedge A(b) \wedge a < b . \rightarrow . B(a) <\rightarrow \mathcal{G}(a,b)$.

    Lemmata 2 and 3 together yield the main lemma:

Lemma 4: $(\exists PR) \lceil \mathcal{L}(P,R) \wedge \mathcal{L}(P,R) \rceil \rightarrow (\exists Q)^\omega . (\forall y)(\forall x)_o^y \lceil Q(x) \wedge Q(y) \rightarrow$
                $\rightarrow \mathcal{G}(x,y) \rceil \vee (\forall y)(\forall x)_o^y \lceil Q(x) \wedge Q(y) \rightarrow \neg \mathcal{G}(x,y) \rceil$.

Proof: Let A,B satisfy $\mathcal{L}(A,B) \wedge \mathcal{L}(A,B)$. Then $(\exists^\omega t)A(t)$ holds by lemma
2.

1. case: $(\exists^\omega t)B(t)$. From lemma 3 and $(\forall t)\lceil B(t) \rightarrow A(t) \rceil$ follows

$(\forall y)(\forall x)_o^y \lceil B(x) \wedge B(y) \rightarrow \mathcal{G}(x,y) \rceil$.

2. case: $(\forall^\omega t)\neg B(t)$. Define a predicate C by

$C(a) <\rightarrow_{df} A(a) \wedge \neg B(a)$.

Then $(\exists^\omega t)C(t)$ holds and as above

$$(\forall y)(\forall x)_o^y \Big[C(x) \wedge C(y) \to \neg\, \zeta(x,y) \Big]. \quad \#$$

Since the conclusion of lemma 4 is just the statement of theorem 2, the proof of theorem 2 is completed by

<u>Lemma 5</u>: $(\exists PR).\ \mathcal{L}(P,R) \wedge \mathcal{L}(P,R)$

which is a consequence of the recursion theorem b.1.

Now we extend theorem 2 to the case of arbitrary many partition sets. At the same time we drop the condition of theorem 1 that the sets of the covering are disjoint. It is easily seen that both versions of theorem 1 are of equal strength, and the new one is much more easy to write down formally.

<u>Theorem 3</u>: $(\forall y)(\forall x)_o^y \bigvee_{i=1}^{m} \zeta_i(x,y) \to$

$$\to (\exists Q)^\omega \bigvee_{i=1}^{m} (\forall y)(\forall x)_o^y \Big[Q(x) \wedge Q(y) \to \zeta_i(x,y) \Big].$$

<u>Proof</u>: By induction over m we get easily the proof of theorem 3 from theorem 2:

The assertion is trivial for $m = 1$; for $m = 2$ it follows from theorem 2 applied to ζ_1, since the premises of theorem 3 give $\neg\,\zeta_1(a,b) \to \zeta_2(a,b)$. So let be $m \geq 3$: Define formulae $\tilde{\zeta}_j$ as ζ_j for $j = 1,\ldots,m-2$ and $\tilde{\zeta}_{m-1}$ as $\zeta_{m-1} \vee \zeta_m$. By induction hypothesis there is an index j, $1 \leq j \leq m-1$ so that

$$(\exists Q)^\omega (\forall y)(\forall x)_o^y \Big[Q(x) \wedge Q(y) \to \tilde{\zeta}_j(x,y) \Big].$$

If $j < m-1$, we are ready. So let $j = m-1$, let D be a predicate so that

$$(\exists^\omega t)D(t) \wedge (\forall y)(\forall x)_o^y \Big[D(x) \wedge D(y) \to \zeta_{m-1}(x,y) \vee \zeta_m(x,y) \Big].$$

Now write down the proof of theorem 2 as before with the following modifications: Change $\mathcal{L}(A,B)$ into

$$\tilde{\mathcal{L}}(A,B) \equiv_{df} (\forall x)\Big[A(x) \leftrightarrow D(x) \wedge (\forall z)_o^x \big[A(z) \to \big[B(z) \leftrightarrow \zeta_{m-1}(z,x) \big] \big] \Big].$$

Similar get $\tilde{\mathcal{L}}$ by replacing the part $x < t \wedge \zeta(x,t)$ in $\mathcal{L}$ by $x < t \wedge D(t) \wedge \zeta(x,t)$, and moreover everywhere ζ by ζ_{m-1}. It is easily seen that these changings amount to a relativization of the quantifiers of $\mathcal{L}$ and $\mathcal{L}$ to the predicate variable D (of course the part $\zeta_{m-1}(a,b)$ will not be relativized). Thus, if we carry through the above proof with $\tilde{\mathcal{L}}$ and $\tilde{\mathcal{L}}$ instead of $\mathcal{L}$ and $\mathcal{L}$ - only using everywhere the argument "$A(a_o)$ holds" instead of "$A(o)$ holds", where a_o is the first element of D -, we get as the consequence of lemmata 4 and 5

$$(\exists Q)^\omega.(\forall x)\Big[Q(x) \to D(x) \Big] \wedge \Big[(\forall y)(\forall x)_o^y \big[Q(x) \wedge Q(y) \to \zeta_{m-1}(x,y) \big] \vee$$
$$\vee\ (\forall y)(\forall x)_o^y \big[Q(x) \wedge Q(y) \to \neg\, \zeta_{m-1}(x,y) \big] \Big].$$

From this and

$$D(a) \wedge D(b) \wedge \neg \zeta_{m-1}(a,b) \rightarrow \zeta_m(a,b)$$

we infer the desired formula

$$(\exists Q)^\omega . (\forall y)(\forall x)_0^y \left[Q(x) \wedge Q(y) \rightarrow \zeta_{m-1}(x,y) \right] \vee$$

$$\vee \ (\forall y)(\forall x)_0^y \left[Q(x) \wedge Q(y) \rightarrow \zeta_m(x,y) \right] . \quad \#$$

At last we extend theorem 3 to the case of arbitrary n-tuples.

__Theorem 4:__ $(\forall x_1,\ldots,x_n) \left[\bigwedge_{j=1}^{n-1} x_j < x_{j+1} \rightarrow \bigvee_{i=1}^{m} \zeta_i(x_1,\ldots,x_n) \right] \rightarrow$

$\rightarrow (\exists Q)^\omega . \bigvee_{i=1}^{m} (\forall x_1,\ldots,x_n) \left[\bigwedge_{j=1}^{n-1} x_j < x_{j+1} \wedge \bigwedge_{j=1}^{n} Q(x_j) \rightarrow \bigvee_{i=1}^{m} \zeta_i(x_1,\ldots,x_n) \right].$

We get theorem 4 from theorem 3 by induction on n, just as we got theorem 3 from theorem 2 by induction on m. Another proof for theorem 4 follows from theorem 1, since SC is complete (theorem 5.a.2, which is proved with the help of theorem 3 only).

d) The normal forms Σ_n^ω.

In the first three sections, we have investigated the system SC by deriving from the axioms some rather strong theorems, thus demonstrating the power of SC. In this section, we continue the investigation of SC from a somewhat opposite point of view. Namely we shall show the weakness of SC by proving that any formula of SC can be brought into a certain handy normal form. This normal form suggests an easily conceivable interpretation (section 2.a), which leads in §3 to the DP. Moreover, it is seen from this interpretation that only a very restricted sort of problem can be formulated in SC. - To get merely the DP, it would suffice to have a normal form for sentences, but for the investigation of definability in SC (Chap.II) we need one for formulae with free variables, too. For the envisaged normal form it makes no difference to admit formulae with free predicate variables, whereas the handling of formulae with free individual variables is better postponed (II.1.c).

So let $\mathfrak{f}(\underline{A}^\ell)$ be a formula containing just the indicated ℓ free predicate variables and no free individual variables. We simplify $\mathfrak{f}$ step by step in the following way:

Step 1: First of all we eliminate superpositions of the successor function by applying one formula of the following lemma 1. Similarly, we make free the arguments of the A_ν and the constant o from the stroke at all; moreover, we eliminate o as argument of the A_ν.

Lemma 1: $\mathfrak{A}(a+n+1) \; :\!\!<\!\!-\!\!>: \; (\exists \underline{x}^w) . \overset{n-1}{\underset{\nu=1}{\wedge}} \; x_\nu' = x_{\nu+1} \wedge x_1 = a' \wedge \mathfrak{A}(x_n')$

$\qquad\qquad \mathfrak{A}(a+n+1) \; :\!\!<\!\!-\!\!>: \; (\forall \underline{x}^w) . \overset{n-1}{\underset{\nu=1}{\wedge}} \; x_\nu' = x_{\nu+1} \wedge x_1 = a' \; -\!\!> \; \mathfrak{A}(x_n')$

Here $\underline{x}^w$ is a string of variables not contained in $\mathfrak{A}(a)$. - We omit the proof.

Step 2: Using the equivalences of the definitions in section a we eliminate the predicate signs $=$, $<$ and $\equiv$ (m). The results of step 1 are not disturbed by step 2.

Step 3: We put the reached formula into prenex normal form.

Step 4: We simplify the prefix by pushing the predicate quantifiers in front in the following way: Only for the moment, we define the rank of a predicate quantifier in the prefix to be the number of individual quantifiers left from it. If not the rank of all predicate quantifiers is zero, let e.g. $(\forall P)$ be the outmost one whose rank is greater than zero. Then the prefix has either the form $..(\forall x)(\forall P)..$ or $..(\exists x)(\forall P)...$

In the first case we commute the both indicated quantifiers; in the
second case we apply the following lemma 2, and get ..$(\exists S)(\forall x)(\forall P)..(\exists z)$.
If we commute again $(\forall x)$ and $(\forall P)$, in both cases the rank of $(\forall P)$ is
diminished. By iterating this procedure we get a formula of the form

$$(\Delta_1)(\Delta_2)\mathfrak{f}_1 \quad,$$

where (Δ_1) is a string of predicate quantifiers, (Δ_2) is a string of individual quantifiers, $\mathfrak{f}_1$ is quantifier-free.

Lemma 2: (Elgot[11],lemma 5.4): Let (Δ) be a string of quantifiers not
containing the variables S and z:

a) $\quad(\exists x)(\Delta)\mathfrak{f}(x) <-> (\exists S)(\forall x)(\Delta)(\exists z).\lceil S(x) -> \mathfrak{f}(x)\rceil \wedge S(z)$

b) $\quad(\forall x)(\Delta)\mathfrak{f}(x) <-> (\forall S)(\exists x)(\Delta)(\forall z).S(z) -> \mathfrak{f}(x) \wedge S(x)$

Proof: The condition on an element can simply be replaced by a condition
on a non-empty set:

$$(\exists x)(\Delta)\mathfrak{f}(x) <-> (\exists S).(\forall x)\lceil S(x) -> (\Delta)\mathfrak{f}(x)\rceil \wedge (\exists z)S(z)$$

Bringing the right side into prenex normal form gives a). b) follows by
negation of both sides. #

Step 5: Whereas in step 4 we have driven out the predicate quantifiers,
we apply now the method of Behmann[1] to drive in the individual quantifiers, using here for the first time the fact that SC has only one-
place predicate variables. By the preceding steps $\mathfrak{f}_1$ is a propositional
formula the prime formulae of which are of the form $\wp(\alpha)$, where $\wp$ is a
free or bound predicate variable and α is a term. Let us assume that
the innermost quantifier of (Δ_2) is of the form $(\forall x)$. We put $\mathfrak{f}_1$ into
conjunctive normal form and use the equivalences

(1) $\quad(\forall x)\lceil\mathcal{O}(x) \wedge \mathcal{L}(x)\rceil <-> (\forall x)\mathcal{O}(x) \wedge (\forall x)\mathcal{L}(x)$

(2) $\quad(\forall x)\lceil\mathcal{O}(x) \vee \mathcal{L}\rceil <-> (\forall x)\mathcal{O}(x) \vee \mathcal{L} \qquad$ (where x is not in $\mathcal{L}$)

to push in the quantifier $(\forall x)$, until it appears with scopes of the form
$\bigvee_i^m \wp_i(x^{(1)})$ only, where $\wp_i$ are predicate variables. If on the other hand
the innermost quantifier of (Δ_2) is existential,we carry through the
dual procedure getting parts $(\exists x) \bigwedge_{i=1}^m \wp_i(x^{(1)})$. Now we repeat this step -
regarding the parts $(\exists x) \bigwedge_{i=1}^m \wp_i(x^{(1)})$ and $(\forall x) \bigvee_i^m \wp_i(x^{(1)})$ as prime formulae.
When (Δ_2) is removed, we bring the resulting formula into disjunctive
normal form getting a disjunction of formulae of the form

$$\ell_i(o) \wedge \bigwedge_{\gamma=1}^{m_i} (\forall t) \mathcal{G}_{i,\gamma}(t) \wedge \bigwedge_{\gamma=1}^{\mathcal{S}_i} (\exists t)\mathcal{h}_{i,\gamma}(t) \quad,$$

where ℓ_i, $\mathcal{G}_{i,\gamma}$ and $\mathcal{h}_{i,\gamma}$ are quantifier-free. By this transformation,
$(\Delta_2)\mathfrak{f}_1$ is brought into what is called "Behmann normal form". Using the

above equivalence (1) in the other direction, in each disjunction we replace the middlest term by $(\forall t) \bigwedge\limits_{\gamma=1}^{m_i} g_{ij}(t)$. If we collect all the steps, we see that we have constructed a formula, equivalent to $(\Delta_2)f_1$, of the form

$$\bigvee_{\iota=1}^{m} \left[\ell_i(o) \wedge (\forall t)\, g_i(t) \wedge \bigwedge_{\gamma=1}^{\ell_i} (\exists t)\, h_{ij}(t) \right],$$

where each ℓ_i, g_i, h_{ij} is quantifier-free, and where ℓ_i contains just the term o and g_i and h_{ij} contain just the terms t and t'.

Step 6: Applying remark a.5 we replace now each conjunction of existential quantifiers by a single $(\exists^\omega t)$ (in section II.1.b it will be seen why we choose $(\exists^\omega t)$ instead of $(\exists t)$); the same we do with single existential quantifiers, using remark a.3. With the help of the equivalence

$$(3) \quad \bigvee_{\iota=1}^{m} (\exists \underline{P}^{\delta})\, \alpha_i(\underline{P}) \;\texttt{<->}\; (\exists \underline{P}^{\ell}) \bigvee_{\iota=1}^{m} \alpha_i(\underline{P})$$

we put the emerging switching predicate quantifiers in front of the whole disjunction, and add the other new parts of the formula to the corresponding old ones. Thus we get

$$(\Delta_3).\ \bigvee_{\iota=1}^{m} \left[\widetilde{\ell}_i(o) \wedge (\forall t)\, \widetilde{g}_i(t) \wedge (\exists^\omega t)\, h_i(t) \right],$$

equivalent to $(\Delta_2)f_1$, where (Δ_3) is a string of existential predicate quantifiers.

Step 7: Using the following lemma 3 we reduce step by step the length m of the disjunction, adding the resulting existential predicate quantifiers to (Δ_3). Thus we have constructed a formula, equivalent to f, of the form

$$(\Delta_1)(\Delta_4).\ \alpha(o) \wedge (\forall t)\, \mathcal{B}(t) \wedge (\exists^\omega t)\, \Gamma(t).$$

Lemma 3: $\bigvee\limits_{\iota=1}^{2} \left[\alpha_i(o) \wedge (\forall t)\, \mathcal{B}_i(t) \wedge (\exists^\omega t)\, \Gamma_i(t) \right]$:<->

<->: $(\exists P).\left\lceil \left\lceil P(o) \wedge \alpha_1(o) \right\rceil \vee \left\lceil \neg P(o) \wedge \alpha_2(o) \right\rceil \right\rceil \wedge$

$\wedge\ (\forall t) \left\lceil \left\lceil P(t) \texttt{<->} P(t') \right\rceil \wedge \left\lceil \left\lceil P(t) \wedge \mathcal{B}_1(t) \right\rceil \vee \left\lceil \neg P(t) \wedge \mathcal{B}_2(t) \right\rceil \right\rceil \right\rceil \wedge$

$\wedge\ (\exists^\omega t) \left\lceil \left\lceil P(t) \wedge \Gamma_1(t) \right\rceil \vee \left\lceil \neg P(t) \wedge \Gamma_2(t) \right\rceil \right\rceil$

The proof is trivial if one keeps in mind the idea of a stable predicate (section a).

Step 8: Let $\underline{R}^{\delta}$ and $\underline{P}^{\dagger}$ be the predicate variables occuring in (Δ_1) and (Δ_4), respectively. If we follow carefully the preceding steps, we see that the formula in step 7 has the form

$$(\Delta_1)(\exists \underline{P}^{\rho}).\ \alpha[\underline{P}(o)] \wedge (\forall t)\, \mathcal{B}[\underline{A}(t),\underline{P}(t),\underline{R}(t),\underline{P}(t'),\underline{R}(t')] \wedge$$
$$\wedge\ (\exists^\omega t)\, \Gamma[\underline{P}(t)]\ .$$

Here the square brackets tell us - as we agreed upon in section a - that the formulae α, $\mathcal{b}$, $\mathcal{c}$ are propositional formulae built up from at most the indicated prime formulae. Further all prime formulae are given with their real arguments: the stroke appears only at the two places where it is indicated. Since we want to have a normal form where the stroke appears only in predicate variables from the innermost string of existential quantifiers, we transform the above formula into

$$(\Delta_1)(\exists \underline{P}^{\#}\underline{Q}^{\hat{s}}) . \alpha[\underline{P}(o)] \wedge (\forall t)\lceil[\underline{R}(t) <-> \underline{Q}(t)] \wedge$$
$$\wedge \; \mathcal{b}\,[\underline{A}(t),\underline{P}(t),\underline{R}(t),\underline{P}(t'),Q(t')]\,\rceil \wedge (\exists^{\omega} t)\mathcal{c}\,[\underline{P}(t)] \; .$$

<u>Definitions:</u> 1) For $\mathcal{n} \geq 1$, we define sets $\underline{\Sigma_\mathcal{n}^\omega}$ and $\underline{\Pi_\mathcal{n}^\omega}$ of formulae by recursion: Σ_1^ω-formulae are of the form

$$(\exists \underline{P}^{\mathcal{m}}) . \alpha[\underline{P}(o)] \wedge (\forall t)\mathcal{b}[\underline{A}^\ell(t),\underline{P}(t),\underline{P}(t')] \wedge (\exists^{\omega} t)\mathcal{c}\,[\underline{P}(t)] \; ,$$

where $\ell \geq o$, $\mathcal{m} > o$. $\Sigma_{\mathcal{m}+1}^\omega$ contains the formulae $(\forall \underline{P}).\mathfrak{f}(\underline{A},\underline{P})$ resp. $(\exists \underline{P}).\mathfrak{f}(\underline{A},\underline{P})$, where $\mathfrak{f}(\underline{A},\underline{B}) \in \Sigma_\mathcal{m}^\omega$, according to whether $\mathcal{n}$ is odd resp. even. $\Pi_\mathcal{n}^\omega$ is defined dually.
2) $\widetilde{\Sigma}_\mathcal{n}^\omega$ is the set of all formulae equivalent to $\Sigma_\mathcal{n}^\omega$-formulae; analogously $\widetilde{\Pi}_\mathcal{n}^\omega$.

Thus e.g. $\Sigma_\mathcal{n}^\omega$-formulae are of the form

$$(\substack{\exists\\\forall}\underline{P}_1^{\ell_1})(\substack{\forall\\\exists}\underline{P}_2^{\ell_2})...(\exists\underline{P}_\mathcal{m}^{\ell_\mathcal{m}}) . \alpha[\underline{P}_\mathcal{n}(o)] \wedge (\forall t)\mathcal{b}[\underline{A}^\ell(t),\underline{P}_1(t),...,\underline{P}_\mathcal{n}(t),\underline{P}_\mathcal{n}(t')] \wedge$$
$$\wedge \; (\exists^{\omega} t)\mathcal{c}\,[\underline{P}_\mathcal{n}(t)] \; ,$$

where $\ell \geq o$, $\ell_1,...,\ell_\mathcal{n} > o$, and the first string of quantifiers is existential or universal according to whether $\mathcal{n}$ is odd or even. - It is very important in handling $\Sigma_\mathcal{n}^\omega$- and $\Pi_\mathcal{n}^\omega$-formulae to have in mind the exact meaning of our use of square brackets, which is recalled in step 8 above. Especially, a $\Sigma_\mathcal{n}^\omega$-formula has no other quantifiers than the predicate quantifiers in front and the two quantifiers $(\forall t)$ and $(\exists^{\omega} t)$ in the kernel.

We have shown by the foregoing procedure (which follows loosely the proof of the analogous lemma 1 in Büchi [2];cf.also Elgot[11]):
<u>Theorem 1</u> (Büchi [3],lemma 11): To any formula $\mathfrak{f}$ not containing free individual variables one can construct effectively an $\mathcal{n} \geq 1$ and a formula $\mathcal{g}$ from $\Sigma_\mathcal{n}^\omega$ such that $\mathfrak{f} <-> \mathcal{g}$ is derivable in SC (thus $\mathfrak{f} \in \widetilde{\Sigma}_\mathcal{n}^\omega$).

Trivially, a Σ_1^ω-formula remains a $\Sigma_\mathcal{n}^\omega$-formula under existential quantification of (a part of) its free predicate variables. Thus one can transform any $\Sigma_\mathcal{n}^\omega$-formula into a Σ_1^ω-formula if one can show that $\widetilde{\Sigma}_\mathcal{n}^\omega$ is closed with respect to negation (simply replace each string $(\forall \underline{P}_i^{\ell_i})$ by $\neg(\exists\underline{P}_i^{\ell_i})\neg$). So we formulate as our

Main problems: 1) To show that $\widetilde{\Sigma}_1^\omega$ is closed under negation.
2) To show that Σ_1^ω-sentences are decidable.

If we can solve these problems, we have a DP for SC. In the next section, we give a natural interpretation of $\widetilde{\Sigma}_1^\omega$-formulae, which makes it possible to attack both problems.

§2. Infinite processes

a) Threads and words

In section 1.a, we have introduced the notion of a "process" to illustrate the meaning of formulae containing free predicate variables. Since any given formula contains only finitely many predicate variables, we have restricted ourselves to (infinite) processes of finitely many elements. This concept – the infinite course-of-value of some finite number of lights or switches – is particularly useful to handle Σ_1^ω-formulae.

So let $\mathfrak{f}(\underline{A}^{\prime\prime})$ be a Σ_1^ω-formula, i.e. a formula of the form

$$(\exists \underline{P}^{\prime\prime}).\, \mathcal{O}[\underline{P}(o)] \,\wedge\, (\forall t)\, \mathcal{L}[\underline{A}^{\prime\prime}(t),\underline{P}(t),\underline{P}(t')] \,\wedge\, (\exists^\omega t)\, \mathcal{L}[\underline{P}(t)] \;.$$

Let any interpretation $\widetilde{\underline{A}}^{\prime\prime}$ of the variables $\underline{A}^{\prime\prime}$ be given. Then the truth of $\mathfrak{f}$ under this interpretation is equivalent to the existence of a process, arising from m lights, which satisfies certain conditions depending on $\mathfrak{f}$ and on $\widetilde{A}$: The initial state of the process has to satisfy the condition $\mathcal{O}$, any two neighbouring states have to satisfy the condition $\mathcal{L}$ (which varies in time depending on $\widetilde{A}$), and moreover there must be infinitely many states which satisfy the condition $\mathcal{L}$. For example, if in the formula

$$\mathcal{G}(A,B) \equiv_{df} (\exists PR).P(o) \wedge \neg R(o) \wedge (\forall t)\lceil \lceil R(t') <-> R(t) \vee A(t)\rceil \wedge$$
$$\wedge \lceil R(t) \wedge B(t) -> \neg P(t)\rceil \rceil \wedge (\exists^\omega t)\lceil R(t) \wedge \neg P(t)\rceil$$

the variable A is interpreted by a predicate $\widetilde{A}$ not everywhere false, and B by any $\widetilde{B}$, then $\mathcal{G}$ is true, and thus via the quantifier $(\exists PR)$ gives rise to a process the second component of which behaves like a switching predicate, whereas the first one is rather arbitrary: it has only to be "off" at infinitely many times, and at each time where $\widetilde{B}$ is true if $\widetilde{A}$ was true before; and it has to be "on" only at the time o.

Definition 1: Let $\mathfrak{f}$ be a Σ_1^ω-formula as above: We call $\mathcal{O}$, $\mathcal{L}$ and $\mathcal{L}$ resp. the initial condition, the transition condition and the final condition of $\mathfrak{f}$. We call the part behind the prefix $(\exists \underline{P}^{\prime\prime})$ the kernel of $\mathfrak{f}$.

Whereas in the preceding section the simple attempt to tidy up the vast mass of formulae has led us to the normal forms Σ_ω^ω and especially to Σ_1^ω, the above interpretation gives another motivation to consider just formulae of this type. It motivates at the same time some steps in the proof of theorem 1.d.1. For example, it is very important that the conditions are propositional formulae and that, moreover, the stroke appears only in the transition condition and only at the arguments of the P_ν, without superpositions; thereby any process satisfying the con-

ditions to a given interpretation $\widetilde{\underline{A}}$, is restricted only by its own state
and that of $\underline{A}$ at the just preceding moment of time. Also the absence of
the free variables in the initial and final condition effects that these
conditions do not depend on $\widetilde{\underline{A}}$; thus the final condition does not vary in
time. The fact that a process corresponding to the predicates satisfying
the predicate quantifiers of a true Σ_1^ω-formula is fixed only by these
simple conditions, and the decidability of SC, are two sides of the same
thing: the seemingly strong (second order!) language of SC can determine
processes only in such a meager way, thus SC-sentences are easily clas-
sified and estimated.

The thinking in processes suggests also the following point of view:
all the time we spoke of the "states" of a given process, i.e. of the
state of all its components in a certain moment of time. Thus it is convenient to regard an n-tuple of predicates as an infinite sequence of
n-tuples of truth values rather than as an n-tuple of infinite sequences
of truth values. We define for $n > o$:

<u>Definition 2:</u> O_n is the set of all n-tuples of truth values, written in
columns. We call those n-tuples <u>n-states</u>. S_n is the set of all ω-sequen-
ces of n-states. We call the elements of $\overline{S_n}$ <u>n-threads.</u>

Thus $O_1 = \{T,F\}$, $\quad O_2 = \left\{ \binom{T}{T}, \binom{T}{F}, \binom{F}{T}, \binom{F}{F} \right\}$, and so on. We use the letters X, Y, Z, if required with indices, to denote n-states for arbitrary
n; if necessary, we add n as an upper index. We extend our convention of
1.a to denote n-tuples, and write $\underline{F}^n$ resp. $\underline{T}^n$ for the elements of O_n
consisting merely of F's resp. T's.

To illustrate S_n, we may think of it as a Peano tree which starts
from one point, and has at every knot 2^n branches coming from; e.g. S_1:

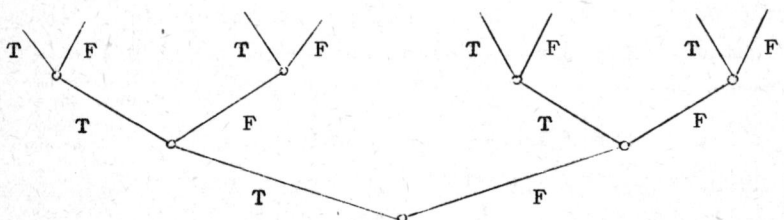

We get any 1-thread as an ascending thread within this tree – therefore
the name. We denote threads by the letters φ, χ, ψ, sets of threads by
the corresponding capitals Φ, X, Ψ. Since any thread φ consists of a
sequence of states, we may speak of its mth state $\varphi(m)$. We indicate suc-
ceeding states of φ by juxtaposing, e.g. $\varphi \equiv \varphi(o)...\varphi(m)X\underline{FF}...\underline{F}Z\underline{TT}...$.

With the help of definition 2 we can make correct our interpretation
of Σ_1^ω-formulae: If we use threads to represent processes, then there is

no difficulty in understanding that a certain process "satisfies a con-
dition"; for any moment of time, the conditions $\mathcal{O}$, $\mathcal{L}$, $\mathcal{L}$ of a Σ_1^ω-for-
mula reduce to truth values if we replace the predicate variables by
the state of a thread at that time. For example, let φ be a 2-thread
satisfying the formula $\mathcal{G}(A,B)$ of p.25, let ψ be a 2-thread which
satisfies together with φ the kernel of $\mathcal{G}$ (i.e. the two components of
φ and of ψ serve as interpretations for the variables A,B,P,R in this
order). Then we may illustrate the situation by the following picture:

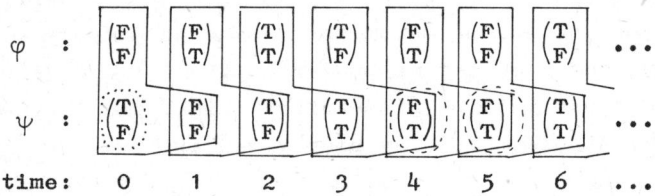

$$\varphi :$$
$$\psi :$$
$$\text{time:} \quad 0 \quad 1 \quad 2 \quad 3 \quad 4 \quad 5 \quad 6 \quad \ldots$$

Here we have framed the triples $\varphi(i)$, $\psi(i)$, $\psi(i+1)$, which satisfy the
transition condition, by continuous lines, similarly $\psi(o)$ and the columns $\psi(i)$ which satisfy the final condition, by a dotted resp. by a
dashed line.

<u>Definition 3:</u> Let a Σ_1^ω-formula $f(\underline{A}^n) \equiv (\exists \underline{P}^m) \mathcal{G}(\underline{A}^n, \underline{P}^m)$ be satisfied by
the n-thread φ. A <u>carrying thread</u> (for f and φ) is an m-thread which
satisfies together with φ the kernel $\mathcal{G}$ of f.

We add a notion useful for the consideration of arbitrary formulae
with free predicate variables. Any SC-formula f containing n free predicate variables, but no individual variables, divides S_n into two
parts, namely the threads which make f true, and the threads which
make f false. We define for those formulae:

<u>Definition 4:</u> The formula $f(\underline{A}^n)$ <u>accepts the n-thread</u> φ if f becomes
true under interpretation of $\underline{A}$ by φ. $\underline{S(f)}$ is the set of all n-threads
accepted by f.

As remarked in section 1.a, in informal proofs we will use free predicate variables to denote predicates. For those cases we carry over
the terminology of threads to that of predicates, and may thus speak
e.g. of an n-<u>predicate</u> $\underline{A}$ <u>accepted by</u> f, of a <u>carrying predicate</u> $\underline{B}$ (for
f and $\underline{A}$), and of its a-th state $\underline{B}(a)$. As in case of threads, usually we
will delete the prefix n-.

Let $\mathcal{G}$ be a Σ_1^ω-sentence with n predicate quantifiers. To look for
whether $\mathcal{G}$ is true or not means to search in S_n for an n-thread satisfying the conditions stated by $\mathcal{G}$. The best way to do so is to build up
threads state by state, as in the picture on p.26, and to pick out at

each knot the "acceptable" directions, i.e. to pick up threads which will be accepted by γ up to this point. Clearly γ is true if and only if this selecting process can be continued at infinity. In fact, this point of view - and thus the thinking in processes at all - is suggested by the paper [27] of Putnam. In this paper, Putnam proves the decidability of a system which is obtained from SC by deleting the quantifiers for predicate variables and by some other minor changes. His procedure is just to construct acceptable threads in the way described above. His main result - which gives the decidability - reads as follows: it is sufficient to examine initial parts of threads up to a certain finite length which depends in an effective manner on the formula considered. If there is an acceptable initial piece under this bound of length, then one gets an acceptable thread by simply repeating a certain part of it; if there is none, then there is no acceptable thread, too. Thus, if a formula is satisfiable at all, then by an ultimately periodic thread.

Literally in the same way we may describe the DP and the result of Büchi. Infinity is expressible in SC merely as ultimately periodic; this makes the decidability of SC. Clearly, SC-formulae are much more complicated than the formulae of the system of Putnam. Therefore the main effort of the DP will be to get a normal form for SC-formulae which is handier than the one of Putnam and makes it thus easy to check the conditions imposed by a formula in this normal form. The above considerations suggest $\Sigma_{\wedge}^{\omega}$ as such a normal form, and indeed, in the next section they will lead us directly to a DP for $\Sigma_{\wedge}^{\omega}$-sentences.

But still the main work has to be done, namely to come from the normal forms $\Sigma_{\wedge\wedge}^{\omega}$ to $\Sigma_{\wedge}^{\omega}$. To this end we need formulae which express the above considerations in connection with the DP of Putnam. I.e. we need formulae which express the following fact: There is a process an initial part of which satisfies the initial condition and the transition condition and - at its end - the final condition of γ. Moreover, we need these formulae also for the case of $\Sigma_{\wedge}^{\omega}$-formulae γ which contain free predicate variables, since we have to look for the satisfiability of such formulae. Thus we are led to the following definition:

<u>Definition 5:</u> Σ^{o} is the set of all SC-formulae of the form

$$(\exists \underline{P}^{m}).\, \mathcal{O}[\underline{P}(a)] \wedge (\forall t)_{a}^{b}\, \mathcal{J}[\underline{A}^{\ell}(t),\underline{P}(t),\underline{P}(t')] \wedge \mathcal{L}[\underline{P}(b)]$$

where ℓ and $m > o$ are arbitrary. $\widetilde{\Sigma^{o}}$ is the set of all formulae equivalent to Σ^{o}-formulae.

As in the case of $\Sigma_{\wedge\wedge}^{\omega}$-formulae one has to be clear on the very structure of Σ^{o}-formulae (cf. the remark behind definitions 1 and 2 in section 1.d). We carry over from Σ_1^ω the concept of <u>initial condition</u>, <u>transition condition</u>, <u>final condition</u> and <u>kernel</u> of a Σ^o-formula.

A Σ^o-formula $f(\underline{A},a,b)$ demands for a process which has to fulfill conditions within a finite intervall of time only. More exactly: A thread φ satisfies the Σ^o-formula $f(\underline{A},r,f)$ if and only if there exists a carrying thread ψ for φ which satisfies the initial condition $\mathcal{O}$ at point r, the final condition $\mathcal{E}$ at point f, and together with φ the transition condition $\mathcal{B}$ between r and f. Therefore the truth of f depends only on the values of φ from r (included) up to f (excluded). (In Trahtenbrot [43], such formulae are called "fictitious outside the interval $(r,f\langle$".) In fact, the formula

$$(\forall t)_a^b \lceil A(t) <-> B(t) \rceil .->. f(\underline{A},a,b) <-> f(\underline{B},a,b),$$

which is a sort of (REXT) (see beginning of section 1.b), is easily derived in SC. Thus we define:

<u>Definition 6</u>: T_n is the set of all finite sequences of n-states, i.e. of elements of O_n. We call the elements of T_n n-words. The <u>length</u> of a word is the number of its states. We include into T_n the empty word $\wedge$ $(= \wedge^n)$.

Thus an n-word arises from juxtaposing a finite number of n-columns of truth values. E.g. $\binom{F}{T}\binom{F}{T}\binom{T}{T}\binom{T}{F}$ is a 2-word of length 4. We use the symbols u, v, w resp. U, V, W (if required with indices) to denote words resp. sets of words. As in the case of threads we indicate by juxtaposing succeeding states and parts of a word u, and write $u(m)$ for the mth state of u; e.g. $u \equiv u(o)...u(m)XvXvZ$. Similarly we compose threads out of words, e.g. $\varphi \equiv uv_1v_2...$. T_n may be represented as the free semigroup generated by the elements of O_n where juxtaposing is the operation, and $\wedge$ is the unit element.

<u>Definition 7</u>: The Σ^o-formula $f(\underline{A}^n,a,b)$ <u>accepts the n-word</u> u of length ℓ iff there exists an n-thread φ such that $\varphi(\iota) \equiv u(\iota)$ for $\iota = o,...,\ell-1$, and $f(\underline{A},a,b)$ becomes true if $\underline{A},a,b$ are interpreted resp. by φ,o,ℓ. In this case we call the word $\psi(o)...\psi(\ell)$ a <u>carrying word</u> (for and u) iff ψ is a carrying thread for φ and f. - $T(f)$ is the set of all n-words accepted by f.

This terminology is useful to make exact the above informal considerations: Let $f(\underline{A}^n)$ be the Σ_1^ω-formula of the very beginning of this section. Let f be satisfied by the n-thread φ, let ψ be a carrying m-thread for φ. Then there exists an infinite sequence $m_1 < m_2 < ...$ of numbers such that $\psi(m_\iota)$ for $\iota = 1,2,...$ satisfies the final condition $\mathcal{E}$. Define $u \equiv_{df} \varphi(o)...\varphi(m_1-1)$, $u_\iota \equiv_{df} \varphi(m_\iota)...\varphi(m_{\iota+1}-1)$ for

$i = 1,2,\ldots,$ and similarly $v \equiv_{df} \psi(o)\ldots\psi(m_1)$, $v_i \equiv_{df} \psi(m_i)\ldots\psi(m_{i+1})$.
Then u is accepted by the Σ°-formula

$$f_1 \equiv_{df} (\exists \underline{P}^m).\, \alpha[\underline{P}(a)] \wedge (\forall t)_a^b\, \mathcal{G}[\underline{A}(t),\underline{P}(t),\underline{P}(t')] \wedge \mathcal{L}[\underline{P}(b)] ,$$

whereas the Σ°-formula

$$f_2 \equiv_{df} (\exists \underline{P}^m).\, \mathcal{L}[\underline{P}(a)] \wedge (\forall t)_a^b\, \mathcal{G}[\underline{A}(t),\underline{P}(t),\underline{P}(t')] \wedge \mathcal{L}[\underline{P}(b)]$$

accepts the words u_i for $i = 1,2,\ldots$. The words v resp. v_i are carrying
words for f_1 and u resp. f_2 and u_i (for any i).

Thus the problem of satisfiability of f has led us to the problem of
satisfiability of certain derived Σ°-formulae. Note that we cannot conclude analogously in the converse direction: For any i, the last state
of v_i is identical with the first one of v_{i+1}; therefore we can splice
the v_i, and get a carrying thread for f. This need not be the case with
the carrying words for an arbitrary sequence $w_1, w_2, \ldots$ from $T(f_2)$.

Before we derive from this consideration a DP for Σ_1^ω (see next section), we introduce still another class of formulae. To this end we recall the above consideration which led us to the definition of Σ°-for-
mulae. There we constructed a carrying thread for a Σ_1^ω-sentence f by
picking out at each point of time acceptable directions, i.e. continuations of the thread which are compatible with the transition condition
of f. This business is particularly easy in the case where at any moment the succeeding state of the carrying thread to be constructed is
determined uniquely by the transition condition from the present state
of the carrying thread (or, if the formula f contains free predicate
variables $\underline{A}$: from the present states of the carrying thread and of the
thread satisfying $f(\underline{A})$). We know those formulae from the discussion on
switching predicates, e.g.

$$(\exists P).\neg P(o) \wedge (\forall t)\big[P(t') <-> P(t) \vee A(t)\big] \wedge (\exists^\omega t)P(t)$$

In general we get those formulae by using recursions of the type (for
notation see section 1.a)

$$\underline{E}^m(o) <-> z^m$$

$$(\forall t)\big[\underline{E}^m(t') <-> \underline{\mathcal{L}}^m[\underline{A}^m(t),\underline{E}(t)]\big]$$

as initial resp. transition condition and adding a final condition.
After a choice for the interpretation of $\underline{A}$, such a recursion determines
uniquely a thread in the tree of m-threads (**above**); thus the truth of
the Σ_1^ω-formula derived from this recursion, only depends on whether the
carrying thread satisfies the final condition infinitely many times. To
such carrying threads correspond "recursive" processes which start from
a fixed beginning state and are steered in their further course by the

information given in form of another process.

It turns out to be very useful to consider those Σ^o-formulae which come from such special Σ_1^ω-formulae:

<u>Definition 8:</u> Σ_R^o is the set of all SC-formulae of the form

$$(\exists \underline{P}^m).\left\lceil \underline{P}(a) <\text{-}> Z^m \right\rceil \wedge (\forall t)_a^b\left\lceil \underline{P}(t') <\text{-}> \underline{L}^m[\underline{A}^\ell(t),\underline{P}(t)] \right\rceil \wedge \lfloor [\underline{P}(b)],$$

where ℓ and $m > o$ are arbitrary. We call these formulae <u>recursive Σ^o-for-</u><u>mulae</u>. $\widetilde{\Sigma}_R^o$ is the set of all formulae equivalent to Σ_R^o-formulae.

Clearly, Σ_R^o-formulae are special Σ^o-formulae. Conversely, which is not so obvious, we will show in section c that every Σ^o-formula is equivalent to a Σ_R^o-formula. This fact will be very useful, since Σ_R^o-formulae are much more handy than Σ^o-formulae.

b) The decidability of Σ^o and Σ_1^ω

Whereas in the last section the consideration of infinite processes has led us from Σ_1^ω via Σ^o to Σ_R^o , now we will go this way back to show step by step the decidability of these classes of formulae, thus solving our main problem 2 (1.d).

We start with the decidability of Σ_R^o-formulae although we will not use this result later on. The reason is that the fact underlying to all the decidability results of this section - namely that the finiteness of O_w for any w gives rise to ultimately periodic threads - is best seen in this special case.

Definition 1: An w-thread φ is <u>ultimately periodic of phase $\not f$ and period $\not z$</u> iff there are w-words u and v of length $\not f$ and $\not z$ respectively such that $\varphi \equiv$ uvvv... . u is the <u>non-periodic</u>, vvv... the <u>periodic part</u> of φ; v is the germ of the periodic part, or shortly the <u>periodic germ</u> of φ. By this definition, if φ is ultimately periodic of phase $\not f$ and period $\not z$, then always $\not z > o$; neither $\not f$ nor $\not z$ are uniquely determined.

Lemma 1: Let $\mathcal{G}$ be the sentence

$$(\exists \underline{P}^m).\left[\underline{P}(o) <-> Z^m\right] \wedge (\forall t)\left[\underline{P}(t') <-> \underline{\mathcal{L}}^m[\underline{P}(t)]\right] .$$

Then $\mathcal{G}$ is true and its (uniquely determined) carrying thread is ultimately periodic with a phase $\not f$ and a period $\not z$ such that $\not f + \not z \le 2^m$.
Proof: We know that $\mathcal{G}$ is true and its carrying-thread is uniquely determined. Let ψ be this carrying thread: ψ is a sequence of elements of O_w. Since O_w contains just 2^m elements, there must be a repetition among the first 2^m+1 states of ψ. Let $\not f$ and $\not z$ be numbers, choosen minimal, such that $o \le \not f < \not z \le 2^m$ and $\psi(\not f) \equiv \psi(\not z)$. Since $\psi(\not z')$ is uniquely determined by $\psi(\not z)$ in the same manner as is $\psi(\not f')$ by $\psi(\not f)$, ψ is of the form uvvv..., where $u \equiv_{df} \psi(o)...\psi(\not f-1)$ and $v \equiv_{df} \psi(\not f)...\psi(\not z-1)$. Choose $\not z =_{df} \not z - \not f$. #

By the following trivial lemmata - which we write down only because we need them later in sections 3.a and II.2.a - lemma 1 gives immediately the decidability of Σ_R^o.

Lemma 2: Let $\not f$ be a Σ^o-formula without free predicate variables: For any w holds $\not f(a,b) <-> \not f(a+w,b+w)$.

The next lemma holds for Σ_R^o-formulae which may contain free predicate variables. It follows from the recursion theorem 1.b.1 and from lemma 1.b.1.

Lemma 3: $(\exists \underline{P}).\left[\underline{P}(o) <-> Z\right] \wedge (\forall t)_o^b\left[\underline{P}(t') <-> \underline{\mathcal{L}}[\underline{A}(t),\underline{P}(t)]\right] \wedge \int[\underline{P}(b)]$

$:<->: (\exists \underline{P}).\left[\underline{P}(o) <-> Z\right] \wedge (\forall t)\left[\underline{P}(t') <-> \underline{\mathcal{L}}[\underline{A}(t),\underline{P}(t)]\right] \wedge \int[\underline{P}(b)]$

Corollary 1: There is an effective procedure to decide on satisfiability of Σ_R^o-formulae not containing free predicate variables. This procedure determines moreover the spectrum of f, i.e. the set of all pairs of numbers which satisfy a given formula of this kind.

Proof: Let $f(a,b)$ be a Σ_R^o-formula of the form

$$(\exists \underline{P}^m).\left[\underline{P}(a) <\!\!-\!\!> Z\right] \wedge (\forall t)_a^b \left[\underline{P}(t') <\!\!-\!\!> \mathcal{L}\left[\underline{P}(t)\right]\right] \wedge \mathcal{S}\left[\underline{P}(b)\right].$$

By lemma 2, it suffices to consider the formula $f(o,b)$. By lemmata 1 and 3, $f(o,b)$ is satisfiable if and only if there is a number $\ell < 2^m$ such that $\mathcal{S}\left[\psi(\ell)\right]$ holds, where ψ is the m-thread defined recursively from the initial and the transition condition of f as in lemma 1. Thus the problem of deciding f is reduced to the propositional calculus. — To determine the spectrum of f we specify the numbers f and z as in the proof of lemma 1, and calculate the numbers $\ell_1, \ldots, \ell_j$ such that $\ell_i < f+z$ and $\mathcal{S}\left[\psi(\ell_i)\right]$ holds for $i = 1, \ldots, j$. Any pair of number satisfying $f(a,b)$ is of the form $(n, n + R_i)$ for some i, $1 \leq i \leq j$, where n is arbitrary and $R_i \equiv \ell_i(z)$. #

We state this last result in another formulation, which we shall need in section II.2.a:

Corollary 2: Let $f(a,b)$ be a Σ_R^o-formula not containing free predicate variables, let f have m predicate quantifiers. Then there are numbers f and z such that $o < z$ and $f + z \leq 2^m$ and

$$a+f \leq b \wedge a+f \leq c \wedge b \equiv c(z) .-\!\!>. f(a,b) <\!\!-\!\!> f(a,c)$$

Now we turn to arbitrary Σ^o-formulae:

Lemma 4: There is an effective procedure to decide for any Σ^o-formula f with free predicate variables whether $T(f) = \emptyset$.

Proof: Let f be a Σ^o-formula:

$$f(\underline{A}^n, a, b) \equiv_{df} (\exists \underline{P}^m).\, \mathcal{O}(a) \wedge (\forall t)_a^b\, \mathcal{L}(t) \wedge \mathcal{S}(b)$$

Let $R =_{df} 2^m$. We will show: If $T(f)$ is not empty, then it contains a word of length not greater than R. Thus suppose $T(f) \neq \emptyset$, let $u \in T(f)$ be an n-word of length $f > R$, let v be the corresponding carrying m-word:

$$\mathcal{O}\left[v(o)\right] \wedge \bigwedge_{i=o}^{f-1} \mathcal{L}\left[u(i), v(i), v(i+1)\right] \wedge \mathcal{S}\left[v(f)\right]$$

Since O_m contains R elements and $f > R$, there must be a repetition among the states of v, say $v(\ell) \equiv v(j)$, where $o \leq \ell < j \leq f$. If we cut out the piece from ℓ to j in both the words u and v, they fulfill again together the conditions:

$$\mathcal{O}\left[v(o)\right] \wedge \bigwedge_{i=o}^{\ell-1} \mathcal{L}\left[u(i), v(i), v(i+1)\right] \wedge \bigwedge_{i=j}^{f-1} \mathcal{L}\left[u(i), v(i), v(i+1)\right] \wedge \mathcal{S}\left[v(f)\right].$$

Thus the word $u(o)\ldots u(\ell-1)u(\acute{\gamma})\ldots u(\acute{\ell}\text{-}1)$, which is shorter than u, lies in $T(\acute{f})$, too. This procedure can be continued, until one reaches a word shorter than $\mathcal{R}+1$.

Thus to show that $T(\acute{f}) \neq \emptyset$ it suffices to consider the finitely many words of length not greater than $\mathcal{P}$. Since, for any fixed word u and any predicate $\underline{A}$ beginning with u, $\acute{f}(\underline{A})$ reduces to a propositional formula, we have the desired result. $\#$

The proof of lemma 4 is carried over nearly unchanged from the proof of theorem 7 in Rabin-Scott[30], which states the decidability of the corresponding "emptiness problem" for finite automata.

For Σ°-formulae with free predicate variables the statement "$T(\acute{f}) \neq \emptyset$" is equivalent to the one "the formula $\acute{f}(\underline{A},a,b) \wedge a \leq b$ is satifiable". We extend the notion of $T(\acute{f})$ to Σ°-formulae without free predicate variables by the definition

$$T(\acute{f}) =_{\mathrm{df}} \left\{\acute{\gamma}; \acute{f}(o,\acute{\gamma}) \text{ holds}\right\} .$$

Then both the above remark and lemma 4 hold for such formulae, too: $T(\acute{f}) \neq \emptyset$ if and only if there exist carrying words whose length is not greater than $\mathcal{R}$. Thus we may reformulate lemma 4 as

Theorem 1: There is an effective procedure to decide on satisfiability of Σ°-formulae. Indeed, if a Σ°-formula $\acute{f}$ with m predicate quantifiers is satisfiable, then $T(\acute{f})$ contains a word of length not greater than 2^{m}.

Remark that $\acute{f}(\underline{A},a,b)$ reduces to a propositional formula, if $b < a$. Thus theorem 1 covers this case, too; at any rate, we will not use this fact.

From theorem 1 we get easily the decidability of Σ_{1}^{ω}:

Theorem 2: (Büchi[3], lemma 12) Σ_{1}^{ω}-sentences are decidable, i.e. there is an effective method for deciding truth of sentences of Σ_{1}^{ω}.

Proof: Let $\mathcal{O}_{f} \equiv (\exists \underline{P}^{m}).\,\mathcal{O}[\underline{P}(o)] \wedge (\forall t)\,\mathcal{L}_{f}[\underline{P}(t),\underline{P}(t')] \wedge (\exists^{\omega}t)\,\mathcal{L}[\underline{P}(t)]$ be in Σ_{1}^{ω}, let $\mathcal{R} =_{\mathrm{df}} 2^{m}$. Assume $\mathcal{O}_{f}$ to be true. Then there exists a carrying m-predicate $\underline{B}$ such that

$$\mathcal{O}[\underline{B}(o)] \wedge (\forall t)\,\mathcal{L}_{f}[\underline{B}(t),\underline{B}(t')] \wedge (\exists^{\omega}t)\,\mathcal{L}[\underline{B}(t)].$$

The **final** condition states that there is an infinite sequence $a_o < a_1 < \ldots$ of numbers such that $\mathcal{L}[\underline{B}(a_{\acute{\iota}})]$ holds for $\acute{\iota} = o,1,\ldots$. Since O_{m} is finite, there must be a repetition among the $\underline{B}(a_{\acute{\iota}})$, say $\underline{B}(a_{\ell}) \equiv \underline{B}(a_{\acute{\jmath}})$ for $\ell < \acute{\jmath} \leq \mathcal{R}$. Clearly, if we replace the remaining part of $\underline{B}$ by simply repeating the part $\underline{B}(a_{\ell}+1)\ldots\underline{B}(a_{\acute{\jmath}})$, the resulting predicate still satisfies the kernel of $\mathcal{O}_{f}$. Thus, if $\mathcal{O}_{f}$ has a carrying predicate, then it has the ultimately periodic one

$$\underline{B}(o)\underline{B}(1)\ldots\underline{B}(a_\ell)\underline{B}(a_\ell+1)\ldots\underline{B}(a_j)\underline{B}(a_\ell+1)\ldots\underline{B}(a_j)\underline{B}(a_\ell+1)\ldots \quad ;$$

the converse is trivial. Define Σ°-formulae ϑ_Y and ℓ_Y for $Y\varepsilon O_m$ by

$$\vartheta_Y(a,b) \equiv_{df} (\exists \underline{P}^m).\,\mathcal{O}[\underline{P}(a)] \wedge (\forall t)_a^b \mathcal{L}[\underline{P}(t),\underline{P}(t')] \wedge \lceil \underline{P}(b) <\!\!-\!\!> Y\rceil ,$$

$$\ell_Y(a,b) \equiv_{df} (\exists \underline{P}^m).\lceil \underline{P}(a) <\!\!-\!\!> Y\rceil \wedge (\forall t)_a^b \mathcal{L}[\underline{P}(t),\underline{P}(t')] \wedge \lceil \underline{P}(b) <\!\!-\!\!> Y\rceil .$$

We have shown: $\mathcal{O}$ is true if and only if there exists an $Y\varepsilon O_m$ such that $\mathcal{L}[Y]$ holds and ϑ_Y and ℓ_Y are satisfiable. Thus the theorem follows from the preceding one. #

If we combine the methods of proof of the two preceeding theorems, we get:

Corollary 3: Σ_1^ω-formulae are decidable, i.e. there is an effective method for deciding satisfiability of Σ_1^ω-formulae.

Proof: We carry through the proof of theorem 2 for a Σ_1^ω-formula $\mathcal{O}(\underline{A})$ instead for a sentence. Let $\underline{A}$ be a predicate satisfying $\mathcal{O}$, let $\underline{B}$ be a carrying predicate for $\mathcal{O}$ and $\underline{A}$, let ℓ and j be as in the proof of theorem 2. If we replace the parts of $\underline{A}$ and $\underline{B}$ above from a_j by repeating the parts of resp. $\underline{A}$ and $\underline{B}$ from $a_\ell+1$ to a_j, the resulting predicates still satisfy the kernel of $\mathcal{O}$. Therefore, if a predicate $\underline{A}$ satisfies $\mathcal{O}$, so does the ultimately periodic one

$$\underline{A}(o)\ldots\underline{A}(a_\ell)\underline{A}(a_\ell+1)\ldots\underline{A}(a_j)\underline{A}(a_\ell+1)\ldots\underline{A}(a_j)\underline{A}(a_\ell+1)\ldots \quad .$$

Thus we conclude: $\mathcal{O}$ is satisfiable if and only if there exists an $Y\varepsilon O_m$ such that $\mathcal{L}[Y]$ holds and ϑ_Y and ℓ_Y are satisfiable. Now corollary 3 follows from theorem 1. #

Corollary 4: If a Σ_1^ω-formula $\mathfrak{f}$ with m predicate quantifiers is satisfiable, then $S(\mathfrak{f})$ contains an ultimately periodic thread φ of phase $\mathfrak{f}$ and period z such that $\mathfrak{f}+z \leq 2^m$. Moreover, there is an ultimately periodic carrying thread for φ and $\mathfrak{f}$ of the same phase and period.

In view of the normal form theorem 3.b.3, corollary 4 shows very earnestly the weakness of SC: We cannot hope to describe in SC processes which are not ultimately periodic (see also II.1.b). Naturally, this weakness rests upon the numbertheoretic weakness of SC: One-place predicate variables and the successor function are the only means to transmit information through the time (cf. 2.a). Thus any process describable in SC can accept information from a fixed finite intervall of time only. Outside this intervall the process can be continued by periodic repetition. – Clearly we could have spared the proof of cor.3, since satisfiability of the Σ_1^ω-formula $\mathcal{O}(\underline{A})$ is equivalent to truth of the Σ_1^ω-sentence $(\exists \underline{R})\,\mathcal{O}(\underline{R})$ (analogously we could have made easier the proof of corollary 1); but

then we would have got a too weak estimation in corollary 4. Especially
in view of the preceding paragraph, it is very important that this estimation does not depend on the number of free predicate variables contained in $\mathcal{G}$: Only the carrying predicate variables transmit information.

c) Boolean operations on Σ^o_R and Σ^o

In the preceding section we have solved our main problem 2 by ascending from Σ^o_R via Σ^o to Σ^ω_1 in showing the decidability. Now we will try the same way in showing the closedness under Boolean operations, thus solving the main problem 1. In this section, we will carry out this plan for Σ^o_R and Σ^o; for Σ^ω_1, we shall not finish it before section 3.b.

For the proof of theorem 1 we need the following trivial lemma, which follows from lemmata 1.b.1+2:

<u>Lemma 1:</u> $(\exists \underline{P}).\lceil \underline{P}(a) \text{ <-> } Z \rceil \wedge (\forall t)^b_a \lceil \underline{P}(t') \text{ <-> } \underline{\mathcal{L}}[\underline{A}(t),\underline{P}(t)] \rceil \wedge \lceil [\underline{P}(b)]$

$:\text{<->}: (\forall \underline{P}).\lceil \underline{P}(a) \text{ <-> } Z \rceil \wedge (\forall t)^b_a \lceil \underline{P}(t') \text{ <-> } \underline{\mathcal{L}}[\underline{A}(t),\underline{P}(t)] \rceil \text{ -> } \lceil [\underline{P}(b)]$

<u>Theorem 1:</u> $\widetilde{\Sigma}^o_R$ is closed under conjunction, disjunction and negation: If $\mathcal{f}(\underline{A}^{\prime\prime},a,b),\ \mathcal{g}(\underline{B}^{\prime\prime\prime},a,b) \varepsilon \widetilde{\Sigma}^o_R$, then $\mathcal{f}(\underline{A},a,b) \wedge \mathcal{g}(\underline{B},a,b),\ \mathcal{f}(\underline{A},a,b) \vee \mathcal{g}(\underline{B},a,b),$ $\neg\mathcal{f}(\underline{A},a,b) \varepsilon \widetilde{\Sigma}^o_R.$

<u>Proof:</u> Let $\mathcal{f}(\underline{A},a,b)$ and $\mathcal{g}(\underline{B},a,b)$ be two Σ^o_R-formulae of the respective forms

$(\exists \underline{P}^*).\lceil \underline{P}(a) \text{ <-> } X \rceil \wedge (\forall t)^b_a \lceil \underline{P}(t') \text{ <-> } \underline{\mathcal{L}}[\underline{A}(t),\underline{P}(t)] \rceil \wedge \lceil [\underline{P}(b)],$

$(\exists \underline{R}').\lceil \underline{R}(a) \text{ <-> } Y \rceil \wedge (\forall t)^b_a \lceil \underline{R}(t') \text{ <-> } \underline{\mathcal{V}}[\underline{B}(t),\underline{R}(t)] \rceil \wedge \mathcal{l} [\underline{R}(b)]$

a) $\mathcal{f}(\underline{A},a,b) \wedge \mathcal{g}(\underline{B},a,b)$ is equivalent to

$(\exists \underline{P}^*\underline{R}').\lceil \underline{P}(a) \text{ <-> } X \rceil \wedge \lceil \underline{R}(a) \text{ <-> } Y \rceil \wedge$

$\wedge (\forall t)^b_a \lceil \lceil \underline{P}(t') \text{ <-> } \underline{\mathcal{L}}[\underline{A}(t),\underline{P}(t)] \rceil \wedge \lceil \underline{R}(t') \text{ <-> } \underline{\mathcal{V}}[\underline{B}(t),\underline{R}(t)] \rceil \rceil \wedge$

$\wedge \lceil [\underline{P}(b)] \wedge \mathcal{l} [\underline{R}(b)] \varepsilon \Sigma^o_R$.

b) $\neg\mathcal{f}(\underline{A},a,b)$ is equivalent to

$\neg(\forall \underline{P}).\lceil \underline{P}(a) \text{ <-> } X \rceil \wedge (\forall t)^b_a \lceil \underline{P}(t') \text{ <-> } \underline{\mathcal{L}}[\underline{A}(t),\underline{P}(t)] \rceil \text{ -> } \lceil [\underline{P}(b)]$

by lemma 1, and thus to

$(\exists \underline{P}).\lceil \underline{P}(a) \text{ <-> } X \rceil \wedge (\forall t)^b_a \lceil \underline{P}(t') \text{ <-> } \underline{\mathcal{L}}[\underline{A}(t),\underline{P}(t)] \rceil \wedge \neg\lceil [\underline{R}(b)] \varepsilon \Sigma^o_R$.

c) The case of disjunction follows from a) and b). #

The above proof reflects the following intuitive fact: A Σ^o_R-formula $\mathcal{f}(\underline{A},a,b)$ determines uniquely a finite piece of a process depending on the thread which is choosen as interpretation of $\underline{A}$. The process "accepts" those threads which make it satisfy the final condition of $\mathcal{f}$. If we have a finite number of Σ^o_R-formulae $\mathcal{f}_i(\underline{A},a,b)$ and its processes, the threads which are accepted by a given propositional function of these processes are just those which make all the processes together satisfy the propositional function of the final conditions of the $\mathcal{f}_i$. - What regards conjunction, these considerations carry over unchanged to arbitrary Σ^o-

formulae, but this is not true for disjunction and negation. So we will show that to any arbitrary Σ°-process there exists an equivalent one (thus accepting the same words) of type Σ°_R, for which again these simple rules of Boolean transformations hold.

__Theorem 2__ (Büchi [3], lemma 2): $\widetilde{\Sigma}^\circ = \widetilde{\Sigma}^\circ_R$, i.e.: to any Σ°-formula one can construct an equivalent Σ°_R-formula.

__Proof:__ Let $\digamma$ be a Σ°-formula:

$$\digamma(\underline{A},a,b) \equiv_{df} (\exists \underline{P}^R) \cdot \alpha[\underline{P}(a)] \wedge (\forall t)^b_a \mathcal{L}[\underline{A}(t),\underline{P}(t),\underline{P}(t')] \wedge \mathcal{L}[\underline{P}(b)]$$

Let $\underline{A}$ be fixed, set $m =_{df} 2^R$ and enumerate the elements of O_R:
$O_R = \{Y_1,\ldots,Y_m\}$. The idea of the proof is to formalize the remark in section a in connection with the DP of Putnam: carrying threads for a Σ°-sentence are paths in O_R, and can be found step by step by picking out at each knot "acceptable" directions, i.e. by looking for words acceptable with respect to the initial and the transition condition of $\digamma$. To this end we define by recursion m predicates which represent at each time the m continuations possible in O_R - whether they are acceptable or not:

$$(1) \qquad \begin{aligned} &D_{\digamma}(o) <-> \alpha[Y_{\digamma}] \\ &D_{\digamma}(a') <-> \bigvee_{\nu=1}^{m}\left[D_{\nu}(a) \wedge \mathcal{L}[\underline{A}(a),Y_{\nu},Y_{\digamma}]\right] \end{aligned} \qquad , \digamma = 1,\ldots,m$$

$D_{\digamma}(\ell)$ holds if and only if there exists a word u of length $\ell+1$ such that $\alpha[u(o)]$ and $\mathcal{L}[\underline{A}(\nu),u(\nu),u(\nu+1)]$ holds for $\nu = o,\ldots,\ell-1$ and $u(\ell)$ is $Y_{\digamma}$. We define $M =_{df} \{\nu; \mathcal{L}[Y_{\nu}]\}$. Then $\digamma(\underline{A},o,\ell)$ is true if and only if $D_{\digamma}(\ell)$ holds for some $\digamma \in M$. Therefore, if we define

$$\alpha_{\digamma} \equiv_{df} \alpha[Y_{\digamma}],$$
$$\mathcal{L}_{\digamma}[\underline{A}(c),\underline{B}^m(c)] \equiv_{df} \bigvee_{\nu=1}^{m}\left[B_{\nu}(c) \wedge \mathcal{L}[\underline{A}(c),Y_{\nu},Y_{\digamma}]\right]$$

for $\digamma = 1,\ldots,m$, we have shown

$$(2) \quad \digamma(\underline{A},a,b) :<->: (\exists \underline{R}^m) \cdot \left[R(a) <-> \underline{\alpha}^m\right] \wedge (\forall t)^b_a\left[R(t') <-> \underline{\mathcal{L}}^m[\underline{A}(t),R(t)]\right] \wedge$$
$$\wedge \bigvee_{\digamma \in M} R_{\digamma}(b)$$

For any $\digamma$, $\alpha[Y_{\digamma}]$ is equivalent to a truth value $Z_{\digamma}$. If we replace in (2) each $\alpha[Y_{\digamma}]$ by the corresponding $Z_{\digamma}$, we get a Σ°_R-formula at the right side. The theorem is proved. #

In [2], corollary 2, Büchi uses the recursion (1) to another purpose. See also the remarks following th.II.1.a.1.- Remark that the equivalence (2) holds only in case $a \leq b$; the same holds for the above lemma 1. This affects neither the informal discussion - since in the definition of $T(\digamma)$ a is always interpreted as being not greater than b - nor later on the formal proofs which use theorem 2 - since there the premise $a \leq b$

is always given.

Let f be a Σ^o-formula, let g be the equivalent Σ^o_a-formula of the above proof, let a predicate $\underline{A}$ satisfying f be fixed: Commonly there will be a lot of carrying threads for f and $\underline{A}$, whereas there is only a single one for g and $\underline{A}$; this single one, however, has a component for each of the states possible for the original carrying threads. Thus we pay for unicity by passing to the powerset (cf.also II.1.a).

Combining theorem 1 and 2 we get

<u>Theorem 3</u> (Büchi [3], lemma 3): $\widetilde{\Sigma}^o$ is closed under conjunction, disjunction and negation: If $f(\underline{A}^w,a,b)$, $g(\underline{B}^w,a,b) \epsilon \widetilde{\Sigma}^o$, then $f(\underline{A},a,b) \wedge g(\underline{B},a,b)$, $f(\underline{A},a,b) \vee g(\underline{B},a,b)$, $\neg f(\underline{A},a,b) \epsilon \widetilde{\Sigma}^o$.

Whereas in the case of $\neg$ we have to pass to the power process, we can avoid theorem 2 in the case of $\wedge$ and $\vee$, thus saving a lot of new quantifiers. Let $f(\underline{A},a,b)$ and $g(\underline{A},a,b)$ be two Σ^o-formulae of the respective form

$$(\exists \underline{P}^\divideontimes). \ \alpha_1[\underline{P}(a)] \ \wedge \ (\forall t)_a^b \ \mathcal{L}_1[\underline{A}(t),\underline{P}(t),\underline{P}(t')] \ \wedge \ \mathcal{L}_1[\underline{P}(b)],$$

$$(\exists \underline{R}^\ell). \ \alpha_2[\underline{R}(a)] \ \wedge \ (\forall t)_a^b \ \mathcal{L}_2[\underline{B}(t),\underline{R}(t),\underline{R}(t')] \ \wedge \ \mathcal{L}_2[\underline{R}(b)] \ .$$

Then $f(\underline{A},a,b) \wedge g(\underline{B},a,b)$ becomes a Σ^o-formula in exactly the same way as in the proof of theorem 1. $f(\underline{A},a,b) \vee g(\underline{B},a,b)$ is equivalent to

$$(\exists \underline{S}^\gamma Q). \lceil \lceil Q(a) \wedge \alpha_1[\underline{S}^\divideontimes(a)] \rceil \vee \lceil \neg Q(a) \wedge \alpha_2[\underline{S}^\ell(a)] \rceil \rceil \wedge$$
$$\wedge \ (\forall t)_a^b \lceil \lceil Q(t) <-> Q(t') \rceil \wedge \lceil \lceil Q(t) \wedge \mathcal{L}_1[\underline{A}(t),\underline{S}^\divideontimes(t),\underline{S}^\divideontimes(t')] \rceil \vee$$
$$\vee \lceil \neg Q(t) \wedge \mathcal{L}_2[\underline{B}(t),\underline{S}^\ell(t),\underline{S}^\ell(t')] \rceil \rceil \rceil \wedge$$
$$\wedge \lceil \lceil Q(b) \wedge \mathcal{L}_1[\underline{S}^\divideontimes(b)] \rceil \vee \lceil \neg Q(b) \wedge \mathcal{L}_2[\underline{S}^\ell(b)] \rceil \rceil .$$

Here $\gamma =_{df} \max(\divideontimes,\ell)$; in the formula, $\underline{S}^\divideontimes$ and $\underline{S}^\ell$ denote the first $\divideontimes$ resp. ℓ of the predicate variables $\underline{S}^\gamma$. The proof of the equivalence starts with the formula

$$(\exists \underline{P}^\divideontimes) \widetilde{f}(\underline{P}) \vee (\exists \underline{R}^\ell) \widetilde{g}(\underline{R}) <-> (\exists \underline{S}^\gamma) \lceil \widetilde{f}(\underline{S}^\divideontimes) \vee \widetilde{g}(\underline{S}^\ell) \rceil \ ,$$

and then uses a stable predicate to file the conditions of f and g to a new Σ^o-type kernel - in the same way as does lemma 1.d.3.

Using that Σ^o is closed under Boolean operations we can extend theorem b.1 to

<u>Corollary 1:</u> There is an effective procedure to decide on validity, on implication, and on equivalence of Σ^o-formulae.

d) Negating Σ_1^ω-formulae

Let $f(\underline{A}^w)$ be a Σ_1^ω-formula of the form

$$(\exists \underline{P}^m).\, \mathfrak{A}[\underline{P}(o)] \,\wedge\, (\forall t)\, \mathfrak{L}[\underline{A}^w(t),\underline{P}(t),\underline{P}(t')] \,\wedge\, (\exists^w t)\, \mathfrak{L}[\underline{P}(t)] \quad.$$

We want to express $\neg f$ as a Σ_1^ω-formula, too. To understand better the
formal construction it is perhaps convenient to give the intuitive
background which leads to the abstract definitions. - We have to get
information about the complementary subsets $S(f)$ and $S(\neg f)$ of S_w. Corollary b.4 tells us that, if $S(f)$ is not empty, it contains an ultimately
periodic thread φ. More exactly: We define formulae ϑ_Y and ℓ_Y for
$Y \varepsilon O_w$ as in the proof of corollary 2.b.3, whose initial condition is
resp. $\mathfrak{A}$ and "<-> Y", whereas the transition condition resp. final condition of both of them are $\mathfrak{L}$ resp. "<-> Y". Then we can represent the
above thread φ in the form uvvv..., where $u \varepsilon T(\vartheta_Y)$ and $v \varepsilon T(\ell_Y)$ for
some Y such that $\mathfrak{L}[Y]$ holds. Clearly, if we replace in φ the word u
by another word $\tilde{u} \varepsilon T(\vartheta_Y)$ (not necessarily of the same length), the resulting thread $\tilde{\varphi}$ satisfies f, too. The same holds if we replace v anywhere by a word $\tilde{v} \varepsilon T(\ell_Y)$. More generally, for any Y so that $\mathfrak{L}[Y]$ holds,
we get threads satisfying f of the form $uv_1v_2v_3...$ where $u \varepsilon T(\vartheta_Y)$ and
$v_i \varepsilon T(\ell_Y)$ for $i = 1,2,...$. Just for the moment, we call a thread ψ
multi-periodic from $\mathfrak{g}$ and $\mathfrak{h}$ if $\mathfrak{g}$ and $\mathfrak{h}$ are Σ^o-formulae and $\psi \equiv$
$uv_1v_2v_3...$ where $u \varepsilon T(\mathfrak{g})$ and $v_i \varepsilon T(\mathfrak{h})$ for $i = 1,2,...$. (This terminology
is not to be confused with Büchi's definition of multi-periodic sets of
words in section II.1.a; both concepts, however, are not independent.)

Thus the proof of corollary b.3 shows that in general there is a lot
of multi-periodic threads satisfying f. But for three reasons the formulae ϑ_Y and ℓ_Y are not yet a good instrument: (1) There can be ultimately periodic threads which satisfy f, but are not multi-periodic
from any pair ϑ_Y, ℓ_Y in this decomposition. (2) Let $\varphi \varepsilon S(f)$ be multi-
periodic from some ϑ_Y, ℓ_Y . If we replace some $v_i \varepsilon T(\ell_Y)$ by some $v \notin T(\ell_Y)$,
the resulting thread need not satisfy $\neg f$. Thus, we cannot use the $\tilde{\Sigma}^o$-
formulae $\neg \vartheta_Y$ and $\neg \ell_Y$ to get multi-periodic threads satisfying $\neg f$.
(3) We do not know anything about threads not multi-periodic. - Thus we
need (perhaps infinitely many) Σ^o-formulae $\mathfrak{h}_i$ such that [1] any ultimately periodic thread is multi-periodic from some $\mathfrak{h}_i$, $\mathfrak{h}_j$; by this we
overcome difficulty (1). Further we need [2] a set M of pairs such that
any thread multi-periodic from some pair $\mathfrak{h}_i$, $\mathfrak{h}_j$ satisfies $\neg f$ if and
only if $(i,j) \varepsilon M$ (difficulty (2)). At least we have to overcome the third
difficulty by showing that [3] any arbitrary thread is multi-periodic
from some $\mathfrak{h}_i$, $\mathfrak{h}_j$.

We get those formulae $\mathfrak{h}_i$ by generalizing the idea of the proof of corollary b.3. Namely let $\underline{A}$ be a predicate such that $\oint(\underline{A})$ is true, let $\underline{C}^{\prime\prime\prime}$ be a carrying predicate for $\underline{A}$. Thus

$$\alpha[\underline{C}(o)] \wedge (\forall t)\, \mathfrak{L}[\underline{A}(t),\underline{C}(t),\underline{C}(t')] \wedge (\exists^{\omega} t)\, \mathfrak{L}[\underline{C}(t)]$$

holds. Let $\mathfrak{k} < \ell$ be any pair of numbers. We replace the part $\underline{A}(\mathfrak{k})\underline{A}(\mathfrak{k}+1)$ ...$\underline{A}(\ell)$ by another word (of arbitrary length, say $\oint$), getting the predicate $\underline{B}$. This operation does not alter the truth value of $\oint$ if there is a carrying predicate for the new part, too, i.e. if

$$(\exists\underline{P}^{\prime\prime\prime}).\left[\underline{P}(\mathfrak{k}) <\!-\!> \underline{C}(\mathfrak{k})\right] \wedge (\forall t)_{\mathfrak{k}}^{\mathfrak{k}+\oint} \mathfrak{L}[\underline{B}(t),\underline{P}(t),\underline{P}(t')] \wedge$$
$$\wedge \left[\underline{P}(\mathfrak{k}+\oint) <\!-\!> \underline{C}(\ell+1)\right]$$

holds. Moreover, any word accepted by this Σ°-formula may replace the considered part of $\underline{A}$. Since we want to exchange infinitely many parts of $\underline{A}$, we have also to ensure that the final condition is satisfied. Thus we define for any pair Y, Z of $O_{\mathcal{W}}$ two formulae

$$\vartheta_{1,Y,Z}(\underline{A},a,b) \equiv_{df} (\exists\underline{P}^{\prime\prime\prime}).\left[\underline{P}(a) <\!-\!> Y\right] \wedge (\forall t)_{a}^{b}\, \mathfrak{L}[\underline{A}(t),\underline{P}(t),\underline{P}(t')] \wedge$$
$$\wedge \left[\underline{P}(b) <\!-\!> Z\right]$$

$$\vartheta_{2,Y,Z}(\underline{A},a,b) \equiv_{df} (\exists\underline{P}^{\prime\prime\prime}).\left[\underline{P}(a) <\!-\!> Y\right] \wedge (\forall t)_{a}^{b}\, \mathfrak{L}[\underline{A}(t),\underline{P}(t),\underline{P}(t')] \wedge$$
$$\wedge \left[\underline{P}(b) <\!-\!> Z\right] \wedge (\exists t)_{a}^{b}\left[\mathfrak{L}[\underline{P}(t)] \vee \mathfrak{L}[\underline{P}(t')]\right]$$

Clearly, $\vartheta_{1,Y,Z}$ is a Σ°-formula. Applying remark 6 of section 1.a we see that also $\vartheta_{2,Y,Z}$ is equivalent to a Σ°-formula, namely to

$$(\exists\underline{P}Q).\left[\underline{P}(a) <\!-\!> Y\right] \wedge \neg Q(a) \wedge$$
$$\wedge (\forall t)_{a}^{b}\left[\mathfrak{L}(t) \wedge \left[Q(t') <\!-\!> Q(t) \vee \mathfrak{L}(t) \vee \mathfrak{L}(t')\right]\right] \wedge$$
$$\wedge \left[\underline{P}(b) <\!-\!> Z\right] \wedge Q(b)$$

Remark that again the equivalence holds only in case a $\leq$ b. By the same reasons as given in case of theorem c.2 above we will not worry about this defect. In the following we will adopt the new formulae as definitions for ϑ_{2}, and thus speak simply of the Σ°-formulae ϑ_{2}.

Now we are able to make precise the above idea of interchanging parts of a given thread. We call two $\mathcal{W}$-words u and v equivalent (with respect to $\oint$) if they are accepted by exactly the same formulae ϑ_{1} and ϑ_{2}, i.e. if for all Y, Z$\varepsilon O_{\mathcal{W}}$:

$$u\varepsilon T(\vartheta_{1,Y,Z}) \text{ if and only if } v\varepsilon T(\vartheta_{1,Y,Z})$$

and

$$u\varepsilon T(\vartheta_{2,Y,Z}) \text{ if and only if } v\varepsilon T(\vartheta_{2,Y,Z}) \quad .$$

Clearly, equivalent words have the wanted property of interchangeability: If the thread $\tilde{\varphi}$ results from φ by replacing any finite part by an equivalent one, then $\varphi \varepsilon S(\mathfrak{f})$ if and only if $\widetilde{\varphi} \varepsilon S(\mathfrak{f})$; we say in this case that φ and $\widetilde{\varphi}$ are equivalent (with respect to $\mathfrak{f}$). - We shall see in the proof of lemma 3.b.2 that we need both types ϑ_1 and ϑ_2, in spite of the fact that $\vartheta_{2,Y,Z}$ implies $\vartheta_{1,Y,Z}$.

The same definition of equivalence will be found in the proof of lemma 9 in Büchi[3]. Indeed, Büchi's "$[Y,u,Z]_{\dot{\iota}}$" says nearly the same as our "$u \varepsilon T(\vartheta_{\dot{\iota},Y,Z})$" ($\dot{\iota} = 1,2$). Büchi then argues as follows: The equivalence is in fact a congruence of words of finite rank (i.e. it is compatible with concatenation of words, and has finitely many equivalence classes). The classes of such a congruence are definable by finite automata (Rabin-Scott[30], theorem 1) and therefore by Σ^o-formulae (Büchi[3], lemma 2).

Since we have defined the equivalence with the help of Σ^o-formulae instead of the automaton-like transition formulae of Büchi, we are able to avoid this roundabout way, and get the wanted Σ^o-formulae directly from the $\vartheta_{\dot{\iota}}$: There are $2 \cdot \rho^2$ formulae $\vartheta_{\dot{\iota}}$, where $\rho =_{df} 2^m$. Any set of formulae $\vartheta_{\dot{\iota}}$ determines the class of words which are accepted by exactly the formulae $\vartheta_{\dot{\iota}}$ of this set, and which are all equivalent. Thus there are at most $\ell =_{df} 2^{2 \cdot \rho^2}$ different equivalence classes. We fix any numeration of the ℓ o-1-sequences of length $2 \cdot \rho^2$, and define for $\dot{\iota} = 1,\dots,\ell$

$$\eta_{\dot{\iota}}(\underline{A},a,b) \equiv_{df} \bigwedge_{Y,Z} (\neg) \vartheta_{1,Y,Z}(\underline{A},a,b) \wedge \bigwedge_{Y,Z} (\neg) \vartheta_{2,Y,Z}(\underline{A},a,b)$$

where the distribution of $\neg$ in $\eta_{\dot{\jmath}}$ coincides with the distribution of zeros in the $\dot{\jmath}$th o-1-sequence. We know (section c) that $\widetilde{\Sigma^o}$ is closed under Boolean operations. Therefore, the $\eta_{\dot{\iota}}$ are $\widetilde{\Sigma^o}$-formulae, too, and we are entitled to speak of $T(\eta_{\dot{\iota}})$. The sets $T(\eta_{\dot{\iota}})$, representing all possible intersections of $T(\vartheta_{\dot{\iota}})$ and their complements, are clearly just the classes of our above equivalence, and thus the congruence classes of Büchi. Notice that the $T(\eta_{\dot{\iota}})$ are disjoint, but clearly may be empty; thus the non-empty sets $T(\eta_{\dot{\iota}})$ form a partition of T_w. The empty set may occur many times. The formal proof is not disturbed by this defect; in the advices for the performance of the DP we will eliminate it (4.a).

It is trivial to derive in SC the property of the $\eta_{\dot{\iota}}$ which corresponds to the covering property of the $T(\eta_{\dot{\iota}})$:

Lemma 1: The formulae $\eta_1,\dots,\eta_\ell$ constructed from the Σ_1^ω-formula $\mathfrak{f}$ are in Σ^o, and it holds:

$$\bigvee_{\dot{\iota}=1}^{\ell} \eta_{\dot{\iota}}(\underline{A},a,b) \quad .$$

Proof: Evidently holds

$$\bigwedge_{Y,Z} \left[\vartheta_{1,Y,Z} \vee \neg \vartheta_{1,Y,Z} \right] \wedge \bigwedge_{Y,Z} \left[\vartheta_{2,Y,Z} \vee \neg \vartheta_{2,Y,Z} \right].$$

The lemma follows by passing to the disjunctive normal form. #

Lemma 1 shows that the formulae η_ν satisfy the requirement[1]. For let $\varphi \equiv uvvv\ldots$ be any ultimately periodic w-thread. Then $u\epsilon T(\eta_\nu)$ and $v\epsilon T(\eta_\jmath)$ for some ν and $\jmath$ by lemma 1; thus φ is multi-periodic from η_ν and $\eta_\jmath$.

For any ν so that $T(\eta_\nu)$ is not empty we choose a fixed word $u_\nu \epsilon T(\eta_\nu)$ (e.g. we could choose the smallest one, where "smallest" is defined in the sense of section 5.b). We define $\varphi_{\nu\jmath} \equiv_{df} u_\nu u_\jmath u_\jmath u_\jmath \ldots$. Then for any $\nu,\jmath$, if φ is multi-periodic from η_ν and $\eta_\jmath$, φ is equivalent with respect to f to the ultimately periodic thread $\varphi_{\nu\jmath}$. Thus we have found a finite number – at most ℓ^2 – of ultimately periodic threads which are representatives for all the threads multi-periodic from the formulae η_ν.

This terminology leads to another interpretation of the theorem of Ramsey if applied to Σ^o-formulae. Clearly, in section 1.d the theorem of Ramsey is intended as a statement on sets of pairs (in general tuples) of numbers. But we have remarked there that for the formal proof the regarded formulae may contain other free variables besides the both marked individual variables. Thus let $\eta_\nu(\underline{A}^w,a,b)$, $\nu = 1,\ldots,m$ be Σ^o-formulae; then theorem 1.d.3 (with a slight modification) reads as follows

(1) $(\forall y)(\forall x)_o^y \bigvee_{\nu=1}^{m} \eta_\nu(\underline{A},x,y) \rightarrow$

$\rightarrow \bigvee_{\nu=1}^{m} (\exists Q)^\omega (\forall y)(\forall x)_o^y \lceil Q(x) \wedge Q(y) \rightarrow \eta_\jmath(\underline{A},x,y) \rceil$

Let φ be any w-thread. Then the formula (1) says: If φ is such that every partial word of φ lies in one of the sets $T(\eta_\nu)$, then there is an index f, $1 \leq f \leq m$, and an infinite sequence $o \leq w_1 < w_2 < \ldots$ so that $\varphi(w_\nu)\ldots\varphi(w_\jmath - 1)\epsilon T(\eta_f)$ for any pair $\nu < \jmath$. This means that φ is multi-periodic from η_y and η_f where y is so that $\varphi(o)\ldots\varphi(w_1 - 1)\epsilon T(\eta_y)$. Moreover, if the $T(\eta_\nu)$ cover T_w, we get the conclusion for any thread φ. Not insisting on a proof of the Ramsey theorem within the system SC we may state this result for arbitrary coverings of T_w:

Corollary 1 (Büchi[3], lemma 1): Let $U_1,\ldots,U_m$ be a covering of T_w, let φ be an w-thread: There exists a number f, $1 \leq f \leq m$ and an infinite sequence $o \leq w_1 < w_2 < \ldots$ so that $\varphi(w_\nu)\ldots\varphi(w_\jmath - 1)\epsilon U_f$ for any pair $\nu < \jmath$.

Now we return to the above discussion, choosing as Σ^o-formulae the congruence formulae η_ν, $\nu = 1,\ldots,\ell$, constructed from the questioned Σ_1^w-formula f. According to lemma 1 the premise of (1) is fulfilled for any $\underline{A}$. Thus any w-thread φ can be represented as $uv_1v_2v_3\ldots$ where, for some pair γ, γ', $u\,\varepsilon\,T(\,\jmath_{\gamma})$ and $v_\iota v_{\iota+1}\ldots v_\jmath\,\varepsilon\,T(\,\jmath_\jmath)$ for every pair $\iota \le \jmath$; especially, φ is multi-periodic from $\jmath_\gamma$ and $\jmath_\jmath$. In this way, the theorem of Ramsey shows that the formulae $\jmath_\iota$ satisfy the above requirement [3]: Any arbitrary w-thread is multiperiodic from some $\jmath_\iota$, $\jmath_\jmath$.

Now the above definition of the threads $\varphi_{\iota,\jmath}$ becomes useful. We may call them the "types of ultimately periodic threads with respect to $\mathfrak{f}$", since any w-thread is equivalent with respect to $\mathfrak{f}$ to at least one of them. In other words: We have found subsets $\Phi_{\iota,\jmath}$ of S_w, $\iota,\jmath = 1,\ldots,\ell$, so that every $\Phi_{\iota,\jmath}$ contains the type $\varphi_{\iota,\jmath}$ and all threads equivalent to it. Thus the $\Phi_{\iota,\jmath}$ need not be disjoint, but any $\Phi_{\iota,\jmath}$ is contained either in $S(\mathfrak{f})$ or in $S(\neg\mathfrak{f})$; thus $S(\mathfrak{f})$ as well as $S(\neg\mathfrak{f})$ is the union of some of the $\Phi_{\iota,\jmath}$. This fact meets our above requirement [2] if we choose as M the set of those pairs $(\iota,\jmath)$ for which $\Phi_{\iota,\jmath} \subseteq S(\neg\mathfrak{f})$ (since then $S(\neg\mathfrak{f}) = \bigcup_{(\iota,\jmath)\varepsilon M} \Phi_{\iota,\jmath}$).

To put these informal considerations into the formal way we define formulae $\mathcal{G}_{\iota,\jmath}(\underline{A}^w)$, $\iota,\jmath = 1,\ldots,\ell$ so that $S(\mathcal{G}_{\iota,\jmath}) = \Phi_{\iota,\jmath}$:

$$\mathcal{G}_{\iota,\jmath}(\underline{A}^w) \equiv_{df} (\exists Q)^\omega\cdot(\exists x)\lceil Q(x) \wedge \jmath_\iota(\underline{A},o,x)\rceil \wedge$$

$$\wedge (\forall y)(\forall x)_o^y\lceil Q(x) \wedge Q(y) \to \jmath_\jmath(\underline{A},x,y)\rceil$$

Then we get immediately from (1) and lemma 1:

<u>Corollary 2:</u> Let $\mathfrak{f}(\underline{A}^w)\varepsilon \Sigma_1^\omega$, let $\mathcal{G}_{\iota,\jmath}$, $\iota,\jmath = 1,\ldots,\ell$, be defined as above:

Then $\quad \overset{\ell}{\underset{\iota,\jmath=1}{\bigvee}} \ \mathcal{G}_{\iota,\jmath}(\underline{A})$.

Corollary 2 says that the sets $\Phi_{\iota,\jmath}$ cover S_w; note that it is here and only here where we use the power of the Ramsey theorem. The property of the $\mathcal{G}_{\iota,\jmath}$ which corresponds to the fact that the $\Phi_{\iota,\jmath}$ do not meet the border line between $S(\mathfrak{f})$ and $S(\neg\mathfrak{f})$ can be stated as

(2) $\quad \mathcal{G}_{\iota,\jmath}(\underline{A}) \wedge \mathcal{G}_{\iota,\jmath}(\underline{B}) \cdot\to\cdot \mathfrak{f}(\underline{A}) \Longleftrightarrow \mathfrak{f}(\underline{B})$, $\iota,\jmath = 1,\ldots,\ell$.

In spite of its intuitive clearness the formula (2) is rather difficult to prove; we shall give a halfformal proof in lemma 3.b.2, and shall complete it to an SC-derivation in section 5.b.

With the help of the formula (2) it is easy to get the wanted set M:

<u>Lemma 2:</u> Let the formulae $\mathcal{G}_{\iota,\jmath}$, $\iota,\jmath = 1,\ldots,\ell$ be constructed from the Σ_1^ω-formula $\mathfrak{f}$ as above. If we define

$$M =_{df} \left\{(\iota,\jmath); (\forall\underline{P})\lceil \mathcal{G}_{\iota,\jmath}(\underline{P}) \to \neg\mathfrak{f}(\underline{P})\rceil\right\},$$

the following equivalence holds:

$$\neg f(\underline{A}) \leftrightarrow \bigvee_{(i,j)\in M} g_{i,j}(\underline{A}).$$

<u>Proof:</u> We note that

(3) $(\forall \underline{P})[g_{i,j}(\underline{P}) \rightarrow f(\underline{P})] \vee (\forall \underline{P})[g_{i,j}(\underline{P}) \rightarrow \neg f(\underline{P})]$, $i,j = 1,\ldots,\ell$

follows directly from the formula (2).

Now in case $(i,j)\varepsilon M$ we get by the definition of M

$$g_{i,j}(\underline{A}) \rightarrow \neg f(\underline{A}) \quad .$$

Therefore the implication holds in one direction:

(4) $\bigvee_{(i,j)\varepsilon M} g_{i,j}(\underline{A}) \rightarrow \neg f(\underline{A})$.

Conversely, suppose $(i,j)\not\varepsilon M$. Thus

$$\neg(\forall \underline{P})[g_{i,j}(\underline{P}) \rightarrow \neg f(\underline{P})] ,$$

which implies by (3)

$$(\forall \underline{P})[g_{i,j}(\underline{P}) \rightarrow f(\underline{P})] .$$

We have proved

$$\bigvee_{(i,j)\notin M} g_{i,j}(\underline{A}) \rightarrow f(\underline{A})$$

or, what amounts to the same,

$$\neg f(\underline{A}) \rightarrow \bigwedge_{(i,j)\notin M} \neg g_{i,j}(\underline{A}) \quad .$$

From corollary 2 follows

$$\bigwedge_{(i,j)\notin M} \neg g_{i,j}(\underline{A}) \rightarrow \bigvee_{(i,j)\in M} g_{i,j}(\underline{A}) \quad .$$

Together we get

$$\neg f(\underline{A}) \rightarrow \bigvee_{(i,j)\in M} g_{i,j}(\underline{A})$$

which - together with (4) - yields the lemma. #

In the next section we will show that the formulae $g_{i,j}$ are in $\widetilde{\Sigma}_1^\omega$. From this we get in section 3.b the closedness of $\widetilde{\Sigma}_1^\omega$ under Boolean operations: Since $\widetilde{\Sigma}_1^\omega$ is closed under disjunction (theorem 3.b.1), the negation of every Σ_1^ω-formula can again be written as a Σ_1^ω-formula. This solves the main problem (1) of 1.d.

Accordingly the "range of questions" is cleared up now: Any conditions imposed on processes in the language of SC can be transformed into the form Σ_1^ω. From corollary 2.b.4 follows - what we have already often maintained - that any set of processes defined in SC contains an ultimately periodic process. Thus the man of practice should try to ask only " Σ_1^ω-questions", and should expect only "ultimately periodic

answers". But about the questions at least one should not think too
bad: also for higher n, a Σ_n^ω- or Π_n^ω-formula has a very clear meaning;
but in general the equivalent Σ_1^ω-formula will have such a lot of predicate quantifiers that it cannot even be written down, much less grasped.
Many problems are e.g. of form Π_2^ω, i.e. "for any process there is a
Σ_1^ω-process ...", but cannot be formulated easily in Σ_1^ω-form. The same
holds for Π_1^ω-formulae which can be read in the convenient form

$$(\forall \underline{P}).\ \alpha(o)\ \wedge\ (\forall t)\ \mathcal{I}(t)\ \rightarrow\ (\forall^\omega t)\ \mathcal{I}(t).$$

For illustration see also the discussion of the similar hierarchy Σ_n,
Π_n in II.1.b.

What regards the "answers" we have maintained in section b that the
weakness of SC becomes evident in the fact that SC can describe only
ultimately periodic processes. Corollary 1, which does not involve SC,
shows that this weakness is rooted in the concept of thread itself: If
$U_1,\ldots,U_m$ is an arbitrary covering of T_n, then any n-process is multi-
periodic from some U_i, U_j, i.e. becomes ultimately periodic by exchang-
ing finite parts exclusively within U_j. The weakness of SC, however,
lies in the fact that to any set of processes definable by an SC-formula f one can effectively construct such a covering with respect to which
$S(f)$ is closed, and can thus describe $S(f)$ and $S(\neg f)$ sufficiently good
by finitely many ultimately periodic processes, namely by the representatives of the covering. Thus, indeed, only relatively simple sets of
processes are definable in SC. - That all these facts are expressible
and derivable within SC, however, shows a certain power of SC - what
regards deriving theorems not what regards defining (sets of) predicates.

§3. The proof of the normal form theorem

a) The periodicity formulae are $\widetilde{\Sigma}_1^\omega$

We want to show that the formulae $\mathcal{O}_{\tilde{\nu}\tilde{\nu}'}$, constructed in the preceding section, are $\widetilde{\Sigma}_1^\omega$-formulae. To this end we need two theorems, which tell us how Σ^o-formulae behave under quantification.

Theorem 1 (Büchi[3], lemma 4): If $\mathfrak{f}(\underline{A},a,b)\varepsilon\widetilde{\Sigma}^o$, then

$$(\exists x)_a^b \mathfrak{f}(\underline{A},x,b), \quad (\forall x)_a^b \mathfrak{f}(\underline{A},x,b)\varepsilon\widetilde{\Sigma}^o$$

Proof: a) Let $\mathfrak{f}(\underline{A},a,b)$ be a Σ^o-formula:

$$\mathfrak{f}(\underline{A},a,b) \equiv_{df} (\exists\underline{P}).\,\mathcal{O}(a) \wedge (\forall t)_a^b \mathcal{L}(t) \wedge \mathcal{L}(b) \quad.$$

Then

$$(\exists x)_a^b \mathfrak{f}(\underline{A},x,b) :\!<\!-\!>: (\exists\underline{P}).\,\mathcal{L}(b) \wedge (\exists x)_a^b \lceil \mathcal{O}(x) \wedge (\forall t)_x^b \mathcal{L}(t)\rceil \quad.$$

Introducing a switching predicate (cf.1.a) we get the equivalent formula

$$(\exists\underline{P}): \mathcal{L}(b) \wedge (\exists Q).Q(a) \wedge \neg Q(b) \wedge (\forall t)_a^b \lceil Q(t') \to Q(t)\rceil \wedge$$
$$\wedge (\forall t)_a^b \lceil Q(t) \wedge \neg Q(t') \to \mathcal{O}(t) \wedge (\forall y)_t^b \mathcal{L}(y)\rceil \quad.$$

Let c be the switching point. Then $a \leq c \leq b \wedge (\forall t)_a^{c'} Q(t) \wedge (\forall t)_c^b \neg Q(t')$. Therefore $(\forall y)_c^b \mathcal{L}(y)$ is the same as $(\forall t)_a^b \lceil Q(t') \to \mathcal{L}(t)\rceil$. If we eliminate in this way the double universal quantification, we get that $(\exists x)_a^b \mathfrak{f}(\underline{A},x,b)$ is equivalent to the Σ^o-formula

$$(\exists\underline{P}Q).Q(a) \wedge (\forall t)_a^b \lceil \lceil Q(t') \to Q(t)\rceil \wedge \lceil Q(t) \wedge \neg Q(t') \to \mathcal{O}(t)\rceil \wedge$$
$$\wedge \lceil \neg Q(t') \to \mathcal{L}(t)\rceil\rceil \wedge \neg Q(b) \wedge \mathcal{L}(b) \quad.$$

Intuitively spoken, a Σ^o-process starting anywhere between a and b can be changed into a Σ^o-process from a to b by adding a switching light which switches on the original process between a and b.
b) The second statement follows from the first by theorem 2.c.3. #

Theorem 2 (Büchi[3], lemma 5): If $\mathfrak{f}(\underline{A},a,b)\varepsilon\widetilde{\Sigma}^o$, then the formulae $(\exists x)\mathfrak{f}(\underline{A},o,x)$ resp. $(\forall x)\mathfrak{f}(\underline{A},o,x)$ can be brought into the form

$$(\exists\underline{P}).\,\mathcal{O}[\underline{P}(o)] \wedge (\forall t)\mathcal{L}[\underline{A}(t),\underline{P}(t),\underline{P}(t')] \wedge (\exists t)\mathcal{L}[\underline{P}(t)]$$

resp.

$$(\exists\underline{P}).\,\mathcal{O}[\underline{P}(o)] \wedge (\forall t)\mathcal{L}[\underline{A}(t),\underline{P}(t),\underline{P}(t')] \quad.$$

Proof: The wanted formulae are of the form Σ_1^ω, except that the final condition has to be satisfied only once in the first case, and is lacking at all in the second case. Thus the theorem is clear for the case of Σ_R^o-formulae, since there the existence of a process depends only on the final condition to be fulfilled. Therefore the existence of a process up to a certain point of time is equivalent to the existence of an unlimited process which satisfies the final condition anywhere; analogously in the second case.

Now, by theorem 2.c.2 we may assume $\mathfrak{f}\epsilon\Sigma_R^o$:

$$\mathfrak{f}(\underline{A},a,b) \equiv_{df} (\exists\underline{P}).\lceil\underline{P}(a) <\text{->} z\rceil \wedge (\forall t)_a^b\lceil\underline{P}(t') <\text{->} \underline{\mathfrak{L}}(t)\rceil \wedge \mathfrak{L}(b) .$$

By lemma 2.b.3 we may omit the recursion bound:

$$\mathfrak{f}(\underline{A},o,c) :<\text{->}: (\exists\underline{P}).\lceil\underline{P}(o) <\text{->} z\rceil \wedge (\forall t)\lceil\underline{P}(t') <\text{->} \underline{\mathfrak{L}}(t)\rceil \wedge \mathfrak{L}(c).$$

If we quantify both sides by $(\exists x)$, and then place the $(\exists x)$ on the right side just before the final condition, we get the first part of the theorem.

To get the second part we transform $\mathfrak{f}(\underline{A},o,c)$, similarly to lemma 2.c.1, into

$$(\forall\underline{P}).\lceil\underline{P}(o) <\text{->} z\rceil \wedge (\forall t)\lceil\underline{P}(t') <\text{->} \underline{\mathfrak{L}}(t)\rceil \text{->} \mathfrak{L}(c) .$$

Thus

$$(\forall x)\mathfrak{f}(\underline{A},o,x) :<\text{->}: (\forall\underline{P}).\lceil\underline{P}(o) <\text{->} z\rceil \wedge (\forall t)\lceil\underline{P}(t') <\text{->} \underline{\mathfrak{L}}(t)\rceil \text{->}$$
$$\text{->} (\forall x)\mathfrak{L}(x) .$$

Returning to existential predicate quantifiers and putting together the universal individual quantifiers we get the desired result

$$(\forall x)\mathfrak{f}(\underline{A},o,x) :<\text{->}: (\exists\underline{P}).\lceil\underline{P}(o) <\text{->} z\rceil \wedge$$
$$\wedge (\forall t)^\lceil\lceil\underline{P}(t') <\text{->} \underline{\mathfrak{L}}(t)\rceil\wedge \mathfrak{L}(t)\rceil . \quad \#$$

Partly we shall use these theorems also for "impure" Σ^o-formulae, i.e. Σ^o-formulae which may contain free predicate variables in the initial and (or) the final condition:

$$(\exists\underline{P}). \alpha[\underline{A}(a),\underline{P}(a)] \wedge (\forall t)_a^b \mathfrak{L}[\underline{A}(t),\underline{P}(t),\underline{P}(t')] \wedge \mathfrak{L}[\underline{A}(b),\underline{P}(b)]$$

We obtain the Σ_R^o-formula equivalent to such a formula by replacing in the proof of theorem 2.c.2 in the formula (2) $\alpha[Y_{\hat{\jmath}}]$ by $\alpha[\underline{A}(a),Y_{\hat{\jmath}}]$ and $\bigvee_{\hat{\jmath}\in M} P_{\hat{\jmath}}(b)$ by $\bigvee_{\hat{\jmath}\in M}\lceil P_{\hat{\jmath}}(b) \wedge \mathfrak{L}[\underline{A}(b),Y_{\hat{\jmath}}]\rceil$. Thus theorem 2.c.2 holds for impure Σ^o-formulae, too, and from this fact follows easily the same extension of theorem 2.c.3 and 1,2 above. Remark that impurity vanishes from the initial condition by application of theorem 1 and from the final condition by application of the second part of theorem 2.

Now we use these facts to get the wanted result on the periodicity formulae $\mathcal{G}_{i,\hat{\jmath}}$ constructed in the preceding section from a given Σ_1^ω-formula $\mathfrak{f}$.

<u>Theorem 3</u> (Büchi[3],p.6): $\mathcal{G}_{i,j} \in \widetilde{\Sigma}_1^\omega$, $i, j = 1, \ldots, l$.

<u>Proof:</u> Let i, j be fixed. We recall the definition of $\mathcal{G}_{i,j}$:

$$(\exists Q).(\exists^\omega t) Q(t) \land (\exists x)\lceil Q(x) \land \hbar_i(\underline{A},o,x)\rceil \land$$

$$\land (\forall y)(\forall x)_o^y \lceil Q(x) \land Q(y) \rightarrow \hbar_j(\underline{A},x,y)\rceil$$

For the following proof we use the remarks on impure Σ^o-formulae stated before theorem 3; we will not mention specially impurity. Compare further section 4.a, where the transformation is written down in detail (with minor changes).

In the middlest part of the formula $\mathcal{G}_{i,j}$ we push $Q(x)$ into the scope of the quantifiers of $\hbar_i$, apply theorem 2, and get:

$$(\exists \underline{P}). \widetilde{\alpha}_i(o) \land (\forall t) \widetilde{\mathcal{B}}_i(t) \land (\exists t) \widetilde{\mathcal{L}}_i(t) \quad .$$

To treat the last part of the formula $\mathcal{G}_{i,j}$ we replace its kernel by

$$\neg Q(x) \lor \neg Q(y) \lor \hbar_j(\underline{A},x,y) \quad .$$

Regarding $\neg Q(x)$ and $\neg Q(y)$ as (extremely impure!) Σ^o-formulae we apply twice theorem 2.c.3 (relative to $\lor$) and get a Σ^o-formula $\widetilde{\hbar}_j$. Applying theorem 1 and then theorem 2 to $\widetilde{\hbar}_j$ we get the formula

$$(\exists \underline{R}). \widehat{\alpha}_i(o) \land (\forall t) \widehat{\mathcal{B}}_i(t) \quad .$$

Now we rearrange, and see that $\mathcal{G}_{i,j}$ is equivalent to

$$(\exists \underline{PRQ}). \alpha_{i,j}(o) \land (\forall t) \mathcal{B}_{i,j}(t) \land (\exists t) \mathcal{L}_{i,j}(t) \land (\exists^\omega t) Q(t) \quad .$$

Application of remark 1.a.7 yields the Σ_1^ω-formula

$$(\exists \underline{PRQS}). \alpha_{i,j}(o) \land \neg S(o) \land$$

$$\land (\forall t) \lceil \mathcal{B}_{i,j}(t) \land \lceil S(t') <-> S(t) \lor \mathcal{L}_{i,j}(t)\rceil\rceil \land$$

$$\land (\exists^\omega t)\lceil S(t) \land Q(t)\rceil \quad . \quad \#$$

b) Boolean operations on Σ_λ^ω-formulae

To show that Σ_λ^ω is closed under negation we need Σ_λ^ω to be closed under both disjunction and conjunction. Thus we cannot dispense with one of the two following theorems.

Theorem 1 (Büchi[3], lemma 7): $\widetilde{\Sigma}_\lambda^\omega$ is closed with respect to disjunction: If $f(\underline{A})$, $g(\underline{B}) \in \widetilde{\Sigma}_\lambda^\omega$, then $f(\underline{A}) \vee g(\underline{B}) \in \widetilde{\Sigma}_\lambda^\omega$.

Proof: By formula (3) and lemma 3 in 1.d. #

Theorem 2 (Büchi[3], lemma 8): $\widetilde{\Sigma}_\lambda^\omega$ is closed with respect to conjunction: If $f(\underline{A})$, $g(\underline{B}) \in \widetilde{\Sigma}_\lambda^\omega$, then $f(\underline{A}) \wedge g(\underline{B}) \in \widetilde{\Sigma}_\lambda^\omega$.

Proof: Let $f(\underline{A}) \equiv_{df} (\exists \underline{P}). \alpha_1(o) \wedge (\forall t)\, \mathcal{b}_1(t) \wedge (\exists^\omega t)\, \mathcal{L}_1(t)$,

$$g(\underline{B}) \equiv_{df} (\exists \underline{R}). \alpha_2(o) \wedge (\forall t)\, \mathcal{b}_2(t) \wedge (\exists^\omega t)\, \mathcal{L}_2(t).$$

Then $f(\underline{A}) \wedge g(\underline{B})$ is equivalent to

$$(\exists \underline{PR}). \alpha_1(o) \wedge \alpha_2(o) \wedge (\forall t)\lceil \mathcal{b}_1(t) \wedge \mathcal{b}_2(t)\rceil \wedge (\exists^\omega t)\, \mathcal{L}_1(t) \wedge$$
$$\wedge (\exists^\omega t)\, \mathcal{L}_2(t) \; .$$

Applying the following lemma 1 we put the two $(\exists^\omega t)$ together to a single one, and get the Σ_λ^ω-formula

$$(\exists \underline{PRQ}). \alpha_1(o) \wedge \alpha_2(o) \wedge \neg Q(o) \wedge (\forall t)\lceil \mathcal{b}_1(t) \wedge \mathcal{b}_2(t) \wedge$$
$$\lceil Q(t') <-> \lceil \neg Q(t) \wedge \mathcal{L}_1(t)\rceil \vee \lceil Q(t) \wedge \neg \mathcal{L}_2(t)\rceil \rceil \rceil \wedge$$
$$\wedge (\exists^\omega t)\lceil Q(t) \wedge \mathcal{L}_2(t)\rceil \; . \quad \#$$

Lemma 1 (McNaughton[24]): $\neg H(o) \wedge (\forall t)\lceil H(t') <-> \lceil \neg H(t) \wedge H_1(t)\rceil \vee$
$\vee \lceil H(t) \wedge \neg H_2(t)\rceil \rceil . -> . (\exists^\omega t) H_1(t) \wedge (\exists^\omega t) H_2(t) <->$
$<-> (\exists^\omega t)\lceil H(t) \wedge H_2(t)\rceil \; .$

Proof: Büchi employs the same idea in the proof of his lemma 8, but he uses H instead of $H \wedge H_2$ in the conclusion. McNaughton remarks in his review[24] that this formula is not valid (as a simple example use H_1, H_2, where $(\forall t) H_1(t)$ and $(\forall t)\neg H_2(t)$), and corrects it in the way above. The idea is the following: H starts with F, is switched on by the first T of H_1, remains so until the next T of H_2 turns it off, is switched on by the next T of H_1 again, and so on. If H_2 vanishes at some point for ever, H will never more be turned off. This defect is compensated by replacing $H(t)$ by the formula $H(t) \wedge H_2(t)$, which is true for $a_1 < a_2 < \ldots$ just in case H_1 and later on H_2 have a T between a_ν and $a_{\nu+1}$, $\nu = 1, 2, \ldots$. Now the proof is easy: Assume H to satisfy the premises of the lemma, define a predicate E by $E(a) <-> H(a) \wedge H_2(a)$; we have to show

$$(\exists^\omega t)H_1(t) \wedge (\exists^\omega t)H_2(t) <-> (\exists^\omega t)E(t)$$

1) Suppose $(\exists^\omega t)H_1(t) \wedge (\exists^\omega t)H_2(t)$, let a be a number;

we want to show: $(\exists t)\lceil a < t \wedge E(t)\rceil$.

1. case: $\neg H(a')$.

Choose b so that $b > a \wedge H_1(b) \wedge (\forall t)_a^b \neg H_1(t)$.

Then $H(b') \wedge (\forall t)_a^b \neg H(t')$.

Choose c so that $c > b \wedge H_2(c) \wedge (\forall t)_b^c \neg H_2(t)$.

Then $(\forall t)_b^c H(t')$ and therefore $E(c)$.

2. case: $H(a')$.

From $H_2(a')$ follows $E(a')$ by definition, from $\neg H_2(a')$ follows $E(c)$ as

in the first case.

2) Assume $(\forall^\omega t)\neg H_1(t) \vee (\forall^\omega t)\neg H_2(t)$;

we want to show: $(\forall^\omega t)\neg E(t)$.

1. case: $(\forall^\omega t)\neg H_2(t)$.

Then $(\forall^\omega t)\neg E(t)$ by definition of E .

2. case: $(\exists^\omega t)H_2(t)$.

Then $(\forall^\omega t)\neg H_1(t)$ by our assumption.

Choose a so that $(\forall t)_a \neg H_1(t)$.

Then $(\forall t)_{a'}\lceil H(t') <-> H(t) \wedge \neg H_2(t)\rceil$.

Choose b so that $b > a \wedge H_2(b) \wedge (\forall t)_a^b \neg H_2(t)$.

Then $(\forall t)_b \neg H(t)$, from which follows $(\forall^\omega t)\neg E(t)$. #

    To get theorem 3 we have to prove a lemma about the formulae $\mathcal{G}_{i,j}$
constructed in 2.d from a Σ_1^ω-formula f; the lemma was formulated as
formula (2) at the end of 2.d.

<u>Lemma 2:</u> For any $i, j = 1, \dots, \ell$:

$$\mathcal{G}_{i,j}(\underline{A}) \wedge \mathcal{G}_{i,j}(\underline{B}) .->. f(\underline{A}) <-> f(\underline{B})$$

<u>Proof:</u> Let i, j and predicates $\underline{A}$ and $\underline{B}$ be fixed.

Assume $\mathcal{G}_{i,j}(\underline{A}) \wedge \mathcal{G}_{i,j}(\underline{B})$. By definition of $\mathcal{G}_{i,j}$ there are sequences
$a_1 < a_2 < \dots$ and $b_1 < b_2 < \dots$ so that

(1)
$$h_i(\underline{A},o,a_1) \wedge h_i(\underline{B},o,b_1) \quad \text{and}$$
$$h_j(\underline{A},a_q,a_{q+1}) \wedge h_j(\underline{B},b_q,b_{q+1}) \quad , \quad q = 1,2,\dots$$

Suppose $\int(\underline{A})$. Then there exists a carrying predicate $\underline{C}$ so that

(2) $\alpha[\underline{C}(o)] \wedge (\forall t)\, \mathcal{L}[\underline{A}(t),\underline{C}(t),\underline{C}(t')] \wedge (\exists^\omega t)\, \mathcal{L}[\underline{C}(t)]$.

From (1) and (2) we shall construct a carrying predicate $\underline{D}$ for $\int(\underline{B})$.
By (2), $\underline{C}(o)...\underline{C}(a_1)$ is a carrying word for $\underline{A}(o)...\underline{A}(a_1-1)$. Therefore
$\vartheta_{1,Y,Z}(\underline{A},o,a_1)$ holds, where $Y \equiv_{df} \underline{C}(o)$ and $Z \equiv_{df} \underline{C}(a_1)$ and $\vartheta_{1,Y,Z}$ as
defined in 2.d. Thus by (1) and the definition of $\gamma_{\check{\nu}}$ in 2.d, $\gamma_{\check{\nu}}$ con-
tains the formula $\vartheta_{1,Y,Z}$ without negation sign. Therefore by (1),
$\vartheta_{1,Y,Z}(\underline{B},o,b_1)$ holds, too. If we choose $\underline{D}_o(o)...\underline{D}_o(b_1)$ as a word
carrying $\underline{B}(o)...\underline{B}(b_1-1)$ in $\vartheta_{1,Y,Z}$, $\alpha[\underline{D}_o(o)]$ holds. In the same way
we get words $\underline{D}_\gamma(b_\gamma)...\underline{D}_\gamma(b_{\gamma+1})$ for $\gamma = 1,2,...$, carrying $\underline{B}(b_\gamma)...$
$..\underline{B}(b_{\gamma+1}-1)$, so that $\underline{D}_{\gamma-1}(b_\gamma) \equiv \underline{D}_\gamma(b_\gamma) \equiv \underline{C}(a_\gamma)$. Thus we may splice the
$\underline{D}_\gamma$ together getting a predicate $\underline{D}$ for which

$\alpha[\underline{D}(o)] \wedge (\forall t)\, \mathcal{L}[\underline{B}(t),\underline{D}(t),\underline{D}(t')]$

holds. That this splicing is possible within SC will be shown in 5.b.
By (2), there exist infinitely many numbers a so that $\mathcal{L}[\underline{C}(a)]$
holds. Therefore there are infinitely many γ such that
$\vartheta_{2,\underline{C}(a_\gamma),\underline{C}(a_{\gamma+1})}(\underline{A},a_\gamma,a_{\gamma+1})$ holds. For these γ we get also
$\vartheta_{2,\underline{C}(a_\gamma),\underline{C}(a_{\gamma+1})}(\underline{B},b_\gamma,b_{\gamma+1})$. We conclude that $(\exists^\omega t)\, \mathcal{L}[\underline{D}(t)]$ holds and
therefore $\int(\underline{B})$. The converse implication $\int(\underline{B}) \to \int(\underline{A})$ follows by rea-
sons of symmetry, i.e. from (SP); the lemma is proved. #

   Now we have our main theorem:

Theorem 3 (Büchi[3], lemma 9): $\widetilde{\Sigma}^\omega_\wedge$ is closed with respect to negation:
If $\int(\underline{A})\epsilon\widetilde{\Sigma}^\omega_\wedge$, then $\neg\int(\underline{A})\epsilon\widetilde{\Sigma}^\omega_\wedge$.

Proof: Already done: The preceding lemma completes the proof of lemma
2.d.2, which gives the theorem together with theorem 1 above and theorem a.3. #

§4. Inquiring into the decision procedure

a) The working of the decision procedure

The theorems 1.d.1 and 3.b.3, together with the considerations at the end of section 1.d, yield the wanted result about $\sum_{\wedge}^{\omega}$:

Theorem 1 (Büchi[3], theorem 1): $\sum_{\wedge}^{\omega}$ is a normal form for SC-formulae without free individual variables.

Theorem 1, theorem 2.b.2, and theorem 3.b.2 give the decidability of SC:

Theorem 2 (Büchi[3], theorem 2): Truth of SC-sentences is decidable.

Here theorem 3.b.2 is used to get $\sum_{\wedge}^{\omega}$-formulae $f_{\nu,\gamma}(\underline{A})$ equivalent to $g_{\nu,\gamma}(\underline{A}) \wedge f(\underline{A})$, and then to construct with the help of theorem 2.b.2 the set $M = \{(\nu,\gamma); \neg(\exists \underline{P}) \, f_{\nu,\gamma}(\underline{P})\}$. –

Instead of examining further the theorems and lemmata involved, whether they are effective, – what had to be done to prove theorem 2 – we give now a complete list of the steps of the DP. This will be done for convenience of people interested mainly in the procedure and not in the proof, and shows at the same time the effectiveness asked for.

First we give a survey on the main steps of the DP by the following flow diagram (the numbers refer to the following parts):

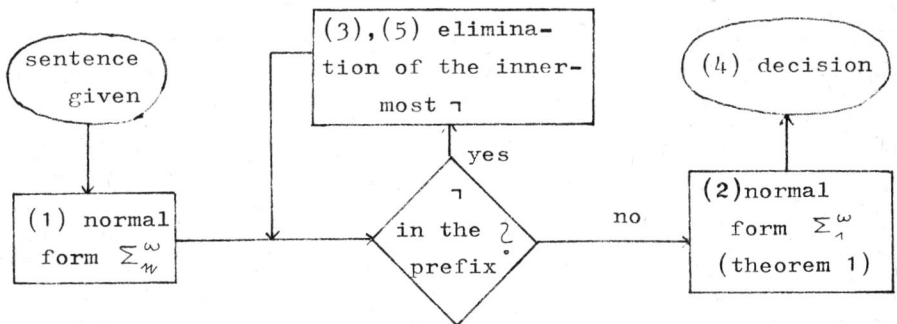

Now we come to the single steps. Let g be any sentence of the system SC:

Part 1 (theorem 1.d.1):

(1) Eliminate superposition of the stroke, and so on (see p.20 step 1).

(2) Eliminate =, <, and ≡ (n) by the definitions in 1.a.

(3) Put into prenex normal form.

(4) Starting with the outermost predicate quantifier shift the predicate quantifiers to the left, simply commuting them with individual quantifiers of the same kind and commuting according to lemma 1.d.2 with indi-

vidual quantifiers of the opposite kind. Get $(\Delta_1)(\Delta_2)\mathfrak{f}_1$, (Δ_1) predicate quantifiers, (Δ_2) individual quantifiers, $\mathfrak{f}_1$ quantifier free.

(5) Bring $(\Delta_2)\mathfrak{f}_1$ into the Behmannsche Normalform, get

$$\overset{\tilde{\tilde{m}}}{\underset{\nu=1}{\bigvee}} \left[\mathfrak{k}_\nu(o) \wedge (\forall t)\, \mathfrak{g}_\nu(t) \wedge \overset{t_\nu}{\underset{\gamma=1}{\bigwedge}} \; (\exists t)\, \mathfrak{h}_{\nu,\gamma}(t) \right],$$

$\mathfrak{k}_\nu,\; \mathfrak{g}_\nu,\; \mathfrak{h}_{\nu,\gamma}$ quantifierfree .

(6) Replace each $\overset{t_\nu}{\underset{\gamma=1}{\bigwedge}}(\exists t)\,\mathfrak{h}_{\nu,\gamma}(t)$ according to remarks 1.a.3+5, and rearrange, getting together with the predicate quantifiers

$$(\Delta_1\underline{R}^\ell)\!:.\; \overset{\tilde{\tilde{m}}}{\underset{\nu=1}{\bigvee}}\;:(\exists\underline{P}\,{}^{t_\nu}_\nu).\; \alpha_\nu(o) \wedge (\forall t)\, \mathfrak{L}_\nu(t) \wedge (\exists^\omega t)\mathfrak{L}_\nu(t) .$$

(7) Apply equivalence (3) and lemma 3 of 1.d, get

$$(\Delta_1\underline{R})(\exists\underline{P}\,{}^{m^*}).\; \alpha^*(o) \wedge (\forall t)\, \mathfrak{L}^*(t) \wedge (\exists^\omega t)\, \mathfrak{L}^*(t) .$$

(8) From this build the formula

$$(\Delta_1\underline{R})(\exists\underline{PQ}).\; \alpha^*(o) \wedge (\forall t)\lceil \mathfrak{L}^*(t) \wedge \lceil\underline{Q}(t) <-> \underline{R}(t)\rceil\rceil \wedge (\exists^\omega t)\, \mathfrak{L}^*(t)$$

where in $\mathfrak{L}^*$ each $R_\nu(t')$ is replaced by $Q_\nu(t')$.

Part 2:

(1) In the thus reached formula

$$(\overset{\exists}{\forall}\underline{P}\,{}^{m_\nu}_1)(\overset{\forall}{\exists}P\,{}^{m_\lambda}_2) \cdots (\forall\underline{P}\,{}^{m_{\kappa-1}}_{n-1})(\exists\underline{P}\,{}^{m^*}_n).\; \alpha[\underline{P}_n(o)] \wedge$$
$$\wedge\; (\forall t)\mathfrak{L}[\underline{P}_1\underline{P}_2\ldots\underline{P}_n(t),\underline{P}_n(t')] \wedge (\exists^\omega t)\mathfrak{L}[\underline{P}_n(t)]$$

replace each $(\forall\underline{P}_\nu)$ by $\neg(\exists\underline{P}_\nu)\neg$.

(2) Drop the predicate quantifiers except $(\exists\underline{P}_n)$ and replace the variables $\underline{P}_1,\ldots,\underline{P}_{n-1}$ within the kernel $\tilde{\mathfrak{f}}_n$ by $\underline{A}_1,\ldots,\underline{A}_{n-1}$.

(3) Go to part 3 and transform $\neg(\exists\underline{P}_n)\,\tilde{\mathfrak{f}}_n(\underline{A}_1,\ldots,\underline{A}_{n-1})$ into a Σ^ω_1-formula $\mathfrak{f}_{n-1}$. Do the same with $\neg(\exists\underline{P}_{n-1})\,\mathfrak{f}_{n-1}(\underline{A}_1,\ldots,\underline{A}_{n-2},\underline{P}_{n-1})$ and so on, until having reached a Σ^ω_1-sentence $\mathfrak{z}$. Go to part 4, get a truth value for $\mathfrak{z}$. This is the truth value of $\mathfrak{Y}$.

Part 3 (theorem 3.b.3): Let $\mathfrak{f}(\underline{A})$ be a Σ^ω_1-formula

$$\mathfrak{f} \equiv_{df} (\exists\underline{P}\,{}^m)\, \alpha(o) \wedge (\forall t)\, \mathfrak{L}(t) \wedge (\exists^\omega t)\, \mathfrak{L}(t)$$

(1) Construct Σ^o-formulae $\vartheta_{\gamma,Y,Z}$, $\gamma = 1,2,Y,Z\epsilon O_m$ as in 2.d.

(2) Construct Σ^o-formulae $\mathfrak{h}_\nu$, $\nu = 1,\ldots,\ell$ as in 2.d.

(3) Construct formulae $\mathfrak{g}_{\nu,\gamma}$, $\nu,\gamma = 1,\ldots,\ell$ as in 2.d.

(4) Go to part 5, transform the $\mathfrak{g}_{\nu,\gamma}$ into Σ^ω_1-formulae.

(5) Using theorem 3.b.2, transform $\mathfrak{g}_{\nu,\gamma}(\underline{A}) \wedge \mathfrak{f}(\underline{A})$ into Σ^ω_1-formulae $\mathfrak{f}_{\nu,\gamma}$.

(6) Go to part 4, decide $(\exists\underline{P})\,\mathfrak{f}_{\nu,\gamma}$, thus constructing the set

$$M =_{df} \{(\check{v},\acute{\jmath}) ; \neg (\exists \underline{P})\, \mathcal{f}_{\check{v},\acute{\jmath}}\}.$$

(7) With the help of theorem 3.b.1 transform $\bigvee_{(\check{v},\acute{\jmath}) \in M} \mathcal{G}_{\check{v},\acute{\jmath}}$ into a $\Sigma_\wedge^{-\omega}$-formula $\mathcal{f}^*$.

(8) Go back to part 2, step (3).

<u>Part 4</u> (theorem 2.b.2): Let $\mathcal{h}$ be a sentence in $\Sigma_\wedge^\omega$:

$$(\exists \underline{P}^m).\, \alpha[\underline{P}(o)] \wedge (\forall t)\, \mathcal{L}[\underline{P}(t),\underline{P}(t')] \wedge (\exists^\omega t)\, \mathcal{L}[\underline{P}(t)]\ .$$

Let $\mathcal{r} =_{df} 2^m$, choose distinct propositional variables $\underline{B}_o^m , \underline{B}_1^m , \ldots, \underline{B}_{\mathcal{r}-1}^m$. Examine the formulae

$$\alpha[\underline{B}_o] \wedge \bigwedge_{\acute{\jmath}=o}^{\ell-1} \mathcal{L}[\underline{B}_{\acute{\jmath}}, \underline{B}_{\acute{\jmath}+1}] \wedge \mathcal{L}[\underline{B}_\ell] \quad \text{and}$$

$$\mathcal{L}[\underline{B}_o] \wedge \bigwedge_{\acute{\jmath}=o}^{\ell-1} \mathcal{L}[\underline{B}_{\acute{\jmath}}, \underline{B}_{\acute{\jmath}+1}] \wedge \mathcal{L}[\underline{B}_\ell] \ , \quad \text{where}$$

$\ell = 1,\ldots,\mathcal{r}-1$, for satisfiability. If there is a pair of words, $u \equiv X_o X_1 \ldots X_\ell$ and $v \equiv Y_o Y_1 \ldots Y_{\tilde{\ell}}$, $\ell, \tilde{\ell} \geq 1$, such that $X_\ell \equiv Y_o \equiv Y_{\tilde{\ell}}$ and u satisfies the relevant among the first formulae and v satisfies the relevant among the second ones, then $\mathcal{h}$ is true; otherwise $\mathcal{h}$ is false.

<u>Part 5</u> (theorem 3.a.3): Let Σ^o-formulae $\mathcal{h}_{\check{v}}$, $\check{v} = 1,\ldots,\ell$ be given.

(1) Apply theorem 2.c.2 to construct Σ_R^o-formulae $\mathcal{h}_{\check{v}}^R$ from $\mathcal{h}_{\check{v}}$, $\check{v} = 1,\ldots,\ell$:

$$\mathcal{h}_{\check{v}}^R(\underline{A},a,b) \equiv_{df} (\exists \underline{P}_{\check{v}}).\, \alpha_{\check{v}}^R(a) \wedge (\forall t)_a^b\, \mathcal{L}_{\check{v}}^R(t) \wedge \mathcal{L}_\wedge^R(b)\ .$$

Let $\check{v},\acute{\jmath}$ be fixed, consider

$$\mathcal{G}_{\check{v},\acute{\jmath}} \equiv_{df} (\exists Q).(\exists^\omega t) Q(t) \wedge (\exists x)\lceil Q(x) \wedge \mathcal{h}_{\check{v}}(\underline{A},o,x)\rceil \wedge$$

$$\wedge (\forall y)(\forall x)_o^y \lceil Q(x) \wedge Q(y) \rightarrow \mathcal{h}_{\acute{\jmath}}(\underline{A},x,y)\rceil\ .$$

(2) Replace the middlest part of $\mathcal{G}_{\check{v},\acute{\jmath}}$ by the formula

$$(\exists \underline{P}_{\check{v}}).\, \alpha_{\check{v}}^R(o) \wedge (\forall t)\, \mathcal{L}_{\check{v}}^R(t) \wedge (\exists x)\lceil Q(x) \wedge \mathcal{L}_{\check{v}}^R(x)\rceil\ .$$

(3) Apply theorem 3.a.2 to the formula

$$(\exists x)_o^y (\exists \underline{P}_{\acute{\jmath}}).\, \alpha_{\acute{\jmath}}^R(x) \wedge Q(x) \wedge (\forall t)_x^y\, \mathcal{L}_{\acute{\jmath}}^R(t) \wedge \neg \mathcal{L}_{\acute{\jmath}}^R(y),$$

get the Σ^o-formula $\tilde{\mathcal{h}}_{\acute{\jmath}}(\underline{A},o,y)$.

(4) With the help of theorem 2.c.2 construct from $\tilde{\mathcal{h}}_{\acute{\jmath}}$ the Σ_R^o-formula

$$\tilde{\mathcal{h}}_{\acute{\jmath}}^R \equiv_{df} (\exists \underline{R}_{\acute{\jmath}}).\, \tilde{\alpha}_{\acute{\jmath}}^R(o) \wedge (\forall t)_o^y\, \tilde{\mathcal{L}}_{\acute{\jmath}}^R(t) \wedge \tilde{\mathcal{L}}_{\acute{\jmath}}^R(y)\ .$$

(5) Replace the last part of $\mathcal{G}_{\check{v},\acute{\jmath}}$ by the formula

$$(\exists \underline{R}_{\acute{\jmath}}).\, \tilde{\alpha}_{\acute{\jmath}}^R(o) \wedge (\forall t)\lceil \tilde{\mathcal{L}}_{\acute{\jmath}}^R(t) \wedge \lceil \neg Q(t) \vee \neg \tilde{\mathcal{L}}_{\acute{\jmath}}^R(t)\rceil\rceil\ .$$

(6) Replace $\mathcal{G}_{\check{v},\acute{\jmath}}$ by the $\Sigma_\wedge^\omega$-formula $\tilde{\mathcal{G}}_{\check{v},\acute{\jmath}}$

$$(\exists \underline{P}_i \underline{R}_j QS). \ \alpha_{\hat{\gamma}}^{\mathbb{R}}(o) \wedge \tilde{\alpha}_{\hat{j}}^{\mathbb{R}}(o) \wedge \neg S(o) \wedge$$

$$\wedge (\forall t) \ulcorner \mathcal{L}_{\nu}^{\mathbb{R}}(t) \wedge \tilde{\mathcal{L}}_{\hat{j}}^{\mathbb{R}}(t) \wedge \ulcorner \neg Q(t) \vee \neg \tilde{\mathcal{L}}_{\hat{j}}^{\mathbb{R}}(t) \urcorner \wedge$$

$$\wedge \ulcorner S(t') <-> S(t) \vee \ulcorner Q(t) \wedge \mathcal{L}_{\nu}^{\mathbb{R}}(t) \urcorner \urcorner \urcorner \wedge (\exists^{\omega} t) \ulcorner S(t) \wedge Q(t) \urcorner \quad .$$

Some improvements and shortenings of the DP will be found in section c.

b) Estimation of growing

A simple look on the DP given in the last section shows that it will never be applicable. The length of the formula questioned grows in such an awful quickness that there will never be a computer able to treat even simple formulae of arbitrary kind. To give an account of this quickness we will now consider the single parts of the DP, estimating each time the rate of growing. That is not to cheer or terrify the reader, but it will demonstrate the most dangerous steps of the DP. In section c) we will then look for shortening of these (and other) steps.

The greatest effort of the DP consists in getting the normal form $\Sigma_\wedge^\omega$, as against which the concluding decision of truth itself seems easy. But the feasibility of part 4, too, depends heavily on the number of predicate quantifiers of the $\Sigma_\wedge^\omega$-sentence resulting from the parts before. If this number is m, one has to decide about the satisfiability of $2(k-1)$ formulae $\mathfrak{h}_\ell^1$, $\mathfrak{h}_\ell^2$, $\ell = 1, \ldots, k-1$, where $k = 2^m$ and each $\mathfrak{h}_\ell^i$ contains $m(\ell+1)$ propositional variables. Thus we shall replace in section c) this very uneconomical procedure by a simpler one (in fact a nearly trivial improvement), but even then part 3 shows that the increasing number of predicate quantifiers is the crucial point of the DP. – Therefore we will now consult the DP step by step looking for upper bounds of the number of predicate quantifiers. To this end define for any formula $\mathfrak{f}$ $L(\mathfrak{f})$ to be the number of its predicate quantifiers. We start with the sentence $\mathcal{O}\!\!\!/$ of section a.

Part 1: Surely it is easy to determine from $\mathcal{O}\!\!\!/$ the number of predicate quantifiers of the prenex normal form reached in step (3). But the growing rate in (4) depends on the distribution of the four different kind of quantifiers in the prefix. This distribution depends on the propositional structure of $\mathcal{O}\!\!\!/$ and is moreover rather arbitrary, since the prenex normal form is not unique. The best we can hope is to get from the number of prime formulae involved an approximative evaluation of the number of quantifiers after step (4) (cf. Specker-Hodes[39]). Instead of this we will replace in section d) steps (1) – (4) by simplified versions. Therefore we start now with step (5), referring to section a for unexplained notation. Let (Δ_1) be the prefix $(\underset{\exists}{\forall}\underline{P}\,\overset{m_1}{1})\ldots(\forall\underline{P}\,\overset{m_{w-1}}{w-1})(\exists\underline{P}\,\overset{m_w}{w})$, where $m_w = 0$ is allowed. Then in (7) $m^* = \max(\mathfrak{f}_i) + \tilde{m} - 1$, which in (8) is increased by adding $\mathfrak{f} =_{df} m_1 + \ldots + m_{w-1}$.

Part 2:
(1) $m_w^* = m^* + \mathfrak{f} + m_w$.
(2) The resulting $\Sigma_\wedge^\omega$-formula is

$$\mathfrak{f}_{\scriptscriptstyle \mathcal{N}} \equiv_{df} (\exists \underline{P}_{\scriptscriptstyle \mathcal{N}}^{\stackrel{m}{m}\stackrel{*}{}}).\alpha[\underline{P}_{\scriptscriptstyle \mathcal{N}}(o)] \wedge (\forall t)\mathcal{L}[A_1^{\stackrel{m}{1}}(t),\dots,\underline{A}_{\stackrel{m-1}{m-1}}^{\stackrel{m-1}{}}(t),\underline{P}_{\scriptscriptstyle \mathcal{N}}(t),\underline{P}_{\scriptscriptstyle \mathcal{N}}(t')] \wedge$$
$$\wedge (\exists^{\scriptscriptstyle \omega} t)\mathcal{L}[\underline{P}_{\scriptscriptstyle \mathcal{N}}(t)] \ .$$

Part 3: Define $m =_{df} m_{\scriptscriptstyle \mathcal{N}}^*$, $\kappa =_{df} 2^m$, $f =_{df} m_1 + \dots + m_{\scriptscriptstyle \mathcal{N}-1}$:
(1) There are κ^2 Σ^o-formulae of each kind ϑ_1 and ϑ_2, $L(\vartheta_1) = m$,
$L(\vartheta_2) = m + 1$.
(2) $L(\neg \vartheta_1) = \kappa$, $L(\neg \vartheta_2) = 2\kappa$, where $\neg \vartheta_1$, $\neg \vartheta_2 \in \Sigma^o$. Define $\ell =_{df} 2^{2 \cdot \kappa^2}$,
$\mathfrak{f} =_{df} 3\kappa^3$. Then there are ℓ Σ^o-formulae $\mathfrak{h}_{\scriptscriptstyle \nu}$, $L(\mathfrak{h}_{\scriptscriptstyle \nu}) \leq \mathfrak{f}$. Remark that
the estimation of $L(\mathfrak{h}_{\scriptscriptstyle \nu})$ is a rough one, since the real length of $\mathfrak{h}_{\scriptscriptstyle \nu}$
depends on how many of the formulae $\vartheta_{\scriptscriptstyle \mathcal{J}}$ are incorporated with a negation sign. E.g. in the best case $L(\mathfrak{h}_{\scriptscriptstyle \nu}) = \kappa^2(2m+1)$.
(3) The number of formulae $\mathcal{G}_{\scriptscriptstyle \nu,\mathcal{J}}$ is ℓ^2.
(4) According to part 5 below: $L(\mathcal{G}_{\scriptscriptstyle \nu,\mathcal{J}}) \leq \mathcal{y}$, where $\mathcal{y} =_{df} 2^{\mathfrak{f}} + 2^{2^{\mathfrak{f}+1}} + 2$.
(5) $L(\mathcal{G}_{\scriptscriptstyle \nu,\mathcal{J}}(\underline{A}^{\mathfrak{f}}) \wedge \mathfrak{f}(\underline{A}^{\mathfrak{f}})) \leq \mathcal{y} + m + 1$, therefore
   $L((\exists \underline{P}^{\mathfrak{f}}).\mathcal{G}_{\scriptscriptstyle \nu,\mathcal{J}}(\underline{P}) \wedge \mathfrak{f}(\underline{P})) \leq \mathfrak{f} + \mathcal{y} + m + 1$.
(7) M contains at most ℓ^2 pairs, thus $L(\neg \mathfrak{f}_{\scriptscriptstyle \mathcal{N}}) \leq \ell^2(\mathcal{y}+1)-1$. In general:

$$L(\neg \mathfrak{f}_{\scriptscriptstyle \mathcal{J}}) \leq \ell_{\scriptscriptstyle \mathcal{J}}^2(\mathcal{y}_{\scriptscriptstyle \mathcal{J}}+1) - 1 = 2^{2^{2(\widetilde{m}_{\scriptscriptstyle \mathcal{J}}+1)}}(2^{3 \cdot 2^{3^{\widetilde{m}_{\scriptscriptstyle \mathcal{J}}}}} + 2^{2^{3 \cdot 2^{3^{\widetilde{m}_{\scriptscriptstyle \mathcal{J}}}}}+1} +3) - 1 \ ,$$

where $\widetilde{m}_{\scriptscriptstyle \mathcal{J}} =_{df} L(\mathfrak{f}_{\scriptscriptstyle \mathcal{J}})$.
Further $L(\mathfrak{f}_{\scriptscriptstyle \mathcal{J}-1}) = L(\neg \mathfrak{f}_{\scriptscriptstyle \mathcal{J}}) + m_{\scriptscriptstyle \mathcal{J}-1}$.

Part 5: Let $L(\mathfrak{h}_{\scriptscriptstyle \nu}) = \mathfrak{f}$, $\nu = 1,\dots, \ell$.
(1) $L(\mathfrak{h}_{\scriptscriptstyle \nu}^{\kappa}) = 2^{\mathfrak{f}}$.
(3) $L(\widetilde{\mathfrak{h}}_{\scriptscriptstyle \nu}) = 2^{\mathfrak{f}} + 1$.
(4) $L(\widetilde{\mathfrak{h}}_{\scriptscriptstyle \nu}^{\kappa}) = 2^{2^{\mathfrak{f}}+1}$.
(6) $L(\widetilde{\mathcal{G}}_{\scriptscriptstyle \nu,\mathcal{J}}) = 2^{\mathfrak{f}} + 2^{2^{\mathfrak{f}}+1} + 2$.

As already stated above the bound given in part 3 is not the best
possible. On an average, $L(\mathfrak{h}_{\scriptscriptstyle \nu})$ lies in the middle between $\kappa^2(2m+1)$ and
$3\kappa^3$. But improvements of this kind are not of great help: Coming to part
3, for example, with the sentence $(\forall P)(\exists R)\mathfrak{f}_2(P,R)$ the estimation tells
us that the number of predicate quantifiers of the resulting $\Sigma_{\scriptscriptstyle \mathcal{N}}^{\omega}$-sen-
tence does not exceed $\approx 10^{3 \cdot 10^6}$. A factor of, say, 2^{16} is there of no
significance. Thus we have to look in the next section whether part 3
can be improved, and how the number $\mathcal{N}$ of the normal form $\Sigma_{\scriptscriptstyle \mathcal{N}}^{\omega}$ can be
minimized.

c) Improvements of the decision procedure

Part 1: Cancel step (3). Then in (4) the predicate variables have to be brought in front of the formula commuting with the individual variables where they stand. It is most important to get as few as possible alternating quantifiers in the prefix. Therefore try to get out at each time as much quantifiers of the same kind as possible.

It is also very important to minimize the total number of predicate quantifiers. Thus it may often be convenient, first to push in quantifiers to make their scope distinct (i.e. to anticipate parts of step (5) applied to predicate quantifiers), and then to apply formula (3) in 1.d and its dual, to fuse predicate quantifiers (cf. e.g. the proof of theorem 2.c.2). For the same reason it may sometimes be better to bring out quantifiers before applying lemma 1.d.2, thus fetching back the original steps (3) and (4). - This now indicated procedure amounts to mix a good cocktail out of steps (3)-(5). When programming this procedure the best would be to let the machine construct all formulae possible in this way, and then to pick out one with minimal (alternating) quantifiers.

The number of (alternating) predicate quantifiers resulting from step (4) can also be lowered by a skilful strategy in steps (1) and (2). Thus additionally to lemma 1.d.1 and the definitions in 1.a, one may use as suitable the following (or similar)

Lemmata:

1) $\quad \alpha(a+m+1) \; :\!\!<\!\!-\!\!>: \; (\exists \underline{P}^m). \bigwedge\limits_{\nu=1}^{m-1} \lceil \lceil P_\nu(t) <\!\!-\!\!> P_{\nu+1}(t') \rceil \wedge$

$$\wedge \lceil P_m(t) <\!\!-\!\!> \alpha(t') \rceil \rceil \wedge P_1(a')$$

2) $\quad \alpha(a+m+1) \; :\!\!<\!\!-\!\!>: \; (\forall \underline{P}^m). \bigwedge\limits_{\nu=1}^{m-1} \lceil \lceil P_\nu(t) <\!\!-\!\!> P_{\nu+1}(t') \rceil \wedge$

$$\wedge \lceil P_m(t) <\!\!-\!\!> \alpha(t') \rceil \rceil \rightarrow P_1(a')$$

3) $\quad a \leq b \; :\!\!<\!\!-\!\!>: \; (\forall P).P(a) \wedge (\forall t) \lceil P(t) \rightarrow P(t') \rceil \rightarrow P(b)$

4) $\quad a < b \; :\!\!<\!\!-\!\!>: \; (\forall P).P(a') \wedge (\forall t) \lceil P(t) \rightarrow P(t') \rceil \rightarrow P(b)$

5) $\quad a \equiv b \; (m) \; :\!\!<\!\!-\!\!>: \; (\exists P).P(a) \wedge P(b) \wedge \bigwedge\limits_{\nu=1}^{m-1} \neg P(a+\nu) \wedge$

$$\wedge (\forall t) \lceil P(t) <\!\!-\!\!> P(t+m) \rceil$$

In (5) it is not needed that the t_ν, g_ν, $g_{\nu,\gamma}$ are conjunctions of prime formula as demanded for the Behmannsche Normalform. It suffices to separate the individual quantifiers. - In (7) lower the number of introduced stable predicates from $\tilde{m} - 1$ to γ where γ is the least number such that $\tilde{m} \leq 2^\gamma$. The method is illustrated in the following

Lemma 6: $\bigvee\limits_{\nu=1}^{4} \lceil A_{\nu}(a) \wedge B_{\nu}(b) \rceil :<->:$

$(\exists PQ).\lceil \lceil P(a) \wedge Q(a) \wedge A_1(a) \rceil \vee \lceil P(a) \wedge \neg Q(a) \wedge A_2(a) \rceil \vee$

$\vee \lceil \neg P(a) \wedge Q(a) \wedge A_3(a) \rceil \vee \lceil \neg P(a) \wedge \neg Q(a) \wedge A_4(a) \rceil \rceil \wedge$

$\wedge (\forall t) \lceil \lceil P(t) <-> P(t') \rceil \wedge \lceil Q(t) <-> Q(t') \rceil \wedge$

$\wedge \lceil \lceil P(b) \wedge Q(b) \wedge B_1(b) \rceil \vee \lceil P(b) \wedge \neg Q(b) \wedge B_2(b) \rceil \vee$

$\vee \lceil \neg P(b) \wedge Q(b) \wedge B_3(b) \rceil \vee \lceil \neg P(b) \wedge \neg Q(b) \wedge B_4(b) \rceil \rceil$

Naturally, it suffices to perform step (8) for those R_{ν} only where $R_{\nu}(t')$ is really contained in $\mathcal{L}$ or $\mathcal{C}$.

Part 3: At the beginning add step

(o) According to corollary 2.b.3, test f for satisfiability. If it is satisfiable, go to (1). If not, get F as truth value for $\mathcal{G}$.

Next insert step

(1a) Using theorem 2.b.1 and corollary 2.c.1 test each $\vartheta_{\gamma,Y,Z}$ for validity and for satisfiability, cancel the valid and the non-satisfiable ones.

For this will not affect the representation of $\neg f$: Suppose for example $\vartheta_{1,X,Y}$ to be valid, consider any $\mathcal{h}_{\nu}$ built up from the old, full list of ϑ's. 1.case: $\vartheta_{1,X,Z}$ is contained unnegated in $\mathcal{h}_{\nu_o}$. Then $\mathcal{h}_{\nu_o}$ reduces to $\mathcal{h}'_{\nu_o}$ which is built up from the new, reduced list, too. 2.case: $\neg \vartheta_{1,X,Y}$ is contained in $\mathcal{h}_{\nu_o}$; then $\mathcal{h}_{\nu_o}$ reduces to F and so does $\mathcal{G}_{\nu,\gamma}$ for any pair (ν,γ) where $\nu = \nu_o$ or $\gamma = \nu_o$. Since all these pairs lie in M, we get $\bigvee\limits_{(\nu,\gamma) \in M} \mathcal{G}_{\nu,\gamma} <-> \bigvee\limits_{(\nu,\gamma) \in M, \nu,\gamma \neq \nu_o} \mathcal{G}_{\nu,\gamma}$. The same argument carries through for non-satisfiable ϑ's. Therefore the two representations of $\neg f$ according to the new or the old list are equivalent.

By an analogous argument insert

(1b) Use corollary 2.c.1 to decide about equivalence among the formulae ϑ_{γ}. Take exactly one from each group of equivalent formulae.

(1c) Use corollary 2.c.1 to decide about implication among the formulae ϑ_{γ}. Then before transforming the formulae $\mathcal{h}_{\nu}$ into Σ^o with the help of the following lemma, shorten them, resp. cancel non-satisfiable ones. Note that always $\vartheta_{2,X,Y} -> \vartheta_{1,X,Y}$.

Lemma 7: If $\alpha -> \mathcal{b}$, then

$\alpha \wedge \mathcal{b} <-> \alpha$,

$\alpha \wedge \neg \mathcal{b} <-> F$,

$\neg \alpha \wedge \neg \mathcal{b} <-> \neg \mathcal{b}$.

Similarly insert

(2a) Test each $\mathcal{h}_{\nu}$ for satisfiability; cancel non-satisfiable $\mathcal{h}_{\nu}$.

Since the conjunction of Σ_R^o-formulae yields a Σ_R^o-formula without producing new predicate quantifiers and since the formulae γ_{i} are used in the following (part 5) only in the form Σ_R^o, one spares a lot of predicate quantifiers $(2^R + 2^l < 2^{R+l}$ in general) by replacing step (2) by (2') Use theorem 2.c.2 to bring the formulae $\vartheta_{\gamma,X,Y}$ into the form Σ_R^o, construct the formulae γ_{i} from these transformed ϑ's.

Part 4: Choose a fixed numbering $X_1, \ldots, X_R$ of the elements of O_m, construct sets $M_1 =_{df} \{i ; \alpha[X_i]\}$, $M_2 =_{df} \{i ; \Sigma[X_i]\}$, and $N =_{df} \{(i,\gamma) ; \mathcal{L}[X_i, X_\gamma]\}$ by evaluating propositional functions. Construct "admissible trees": take a number from M_1 or from M_2, extend step by step by fitting pairs out of N, if there are any; do not extend further branches which have reached length R.

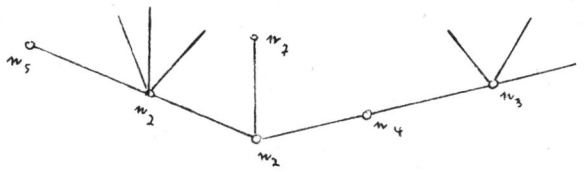

Thus any two neighbouring knots of an admissible tree constitute a pair from N. While constructing look for satisfying sequences: If the procedure yields two sequences so that the first starts in M_1 and terminates in M_2 and the second has this terminal element as initial and final element, then $\mathcal{G}$ is true. Otherwise $\mathcal{G}$ is false.

The estimation in section c shows that the length of a formula reaching part 4 is in general very large. Thus the performability of part 4 depends on having a very good computer-DP for propositional formulae. The one available for us (written for the Siemens computer 2002 by Prof. O.Herrmann, University of Heidelberg) is able to work on formulae containing up to 200 different propositional variables. This number may be enlarged by writing the same program for a bigger and faster computer; but clearly only a program would help capable of 10^4 or 10^5 or 10^6 propositional variables.

Part 5: The relation between the advices in section b and the proof of theorem 3.a.3 ist best seen by replacing in $\mathcal{G}_{i,\gamma}$ the partial formula

$$(\forall y)(\forall x)_o^y \lceil Q(x) \wedge Q(y) \to \gamma_{\gamma}(\underline{A}, x, y) \rceil$$

by the equivalent one

$$\neg(\exists y)\lceil Q(y) \wedge (\exists x)_o^y \lceil Q(x) \wedge \neg \gamma_{\gamma}(\underline{A}, x, y) \rceil \rceil ;$$

steps (3) –(5) act just on this formula.

The transformations of steps (3) –(5) create the most new quantifiers of the whole procedure of theorem 3.b.3 (cf. section b. $2^{2^{\int}}$ is a number inaccessible for machines already for very small $\int$). Thus let us look **for a way out.**

To this end compare the formulae $\mathcal{G}_{\hat{\nu},\hat{\gamma}}$ with the following formulae $\overline{\mathcal{G}}_{\hat{\nu},\hat{\gamma}}$:

$$(\exists \underline{P}^{\delta}\underline{R}^{\delta}QS).S(o) \wedge \alpha_{\nu}[\underline{P}(o)] \wedge (\forall t)\lceil \lceil S(t') \rightarrow S(t)\rceil \wedge$$

$$\wedge \lceil S(t') \rightarrow \mathcal{B}_{\nu}[\underline{A}(t),\underline{P}(t),\underline{P}(t')]\rceil \wedge$$

$$\wedge \lceil S(t) \wedge \neg S(t') \rightarrow \Gamma_{\nu}[\underline{P}(t)] \wedge \alpha_{\gamma}[\underline{R}(t)] \wedge Q(t)\rceil \wedge$$

$$\wedge \lceil Q(t) \wedge \neg S(t') \rightarrow \mathcal{B}_{\gamma}[\underline{A}(t),\underline{R}(t),\underline{R}(t')]\rceil \wedge$$

$$\wedge \lceil \neg Q(t) \wedge \neg S(t) \rightarrow \mathcal{B}_{\gamma}[\underline{A}(t),\underline{P}(t),\underline{P}(t')]\rceil \wedge$$

$$\wedge \lceil Q(t) \wedge \neg Q(t') \wedge \neg S(t') \rightarrow \Gamma_{\gamma}[\underline{R}(t')] \wedge \alpha_{\gamma}[\underline{P}(t')]\rceil \wedge$$

$$\wedge \lceil \neg Q(t) \wedge Q(t') \wedge \neg S(t) \rightarrow \Gamma_{\gamma}[\underline{P}(t')] \wedge \alpha_{\gamma}[\underline{R}(t')]\rceil\rceil \wedge$$

$$\wedge (\exists^{\omega}t)\lceil Q(t) \wedge \neg S(t)\rceil \wedge (\exists^{\omega}t)\neg Q(t) \quad ,$$

where the involved formulae α_{ν} etc. come from the formulae $\mathcal{G}_{\nu}$, $\mathcal{G}_{\gamma}$ and where $\int$ is the length of $\mathcal{G}_{\nu}$, $\mathcal{G}_{\gamma}$. It is easy to see that, whereas $\mathcal{G}_{\hat{\nu},\hat{\gamma}}$ says intuitively "there is an infinite sequence $a_1 < a_2 < \ldots$ such that (1) $\mathcal{G}_{\nu}(o,a_1)$ and (2) $\mathcal{G}_{\gamma}(a_{\int},a_{\gamma})$ for any $\int < \gamma$ ", $\overline{\mathcal{G}}_{\hat{\nu},\hat{\gamma}}$ has the somewhat weaker meaning arising from this by replacing (2) by "(2') $\mathcal{G}_{\gamma}(a_{\gamma},a_{\gamma+1})$ for $\gamma = 1,2,\ldots$". We have replaced simply the pulsating predicate of $\mathcal{G}_{\hat{\nu},\hat{\gamma}}$ [satisfying($\exists Q$)], the T's of which determine the points a_{γ}, by two switching predicates [satisfying($\exists QS$)], the second of which switches only once, thus determining the beginning piece, whereas the first one is a switching predicate in a generalized sense; namely it switches infinitely many times, being true from $a_{2\gamma-1}$ up to $a_{2\gamma}$ for $\gamma = 1,2,\ldots$, and false otherwise. This enables us to distinguish between odd and even indices of the a_{γ} and thus to use simply two $\int$-predicates alternatly as carrying predicates for $\mathcal{G}_{\gamma}$.

Using the above intuitive translation it is easy to see that $\mathcal{G}_{\hat{\nu},\hat{\gamma}}$ implies $\overline{\mathcal{G}}_{\hat{\nu},\hat{\gamma}}$. Thus corollary 2.d.2 holds for $\overline{\mathcal{G}}_{\hat{\nu},\hat{\gamma}}$ instead of $\mathcal{G}_{\hat{\nu},\hat{\gamma}}$, too. Further $\overline{\mathcal{G}}_{\hat{\nu},\hat{\gamma}}$ suffices to derive lemma 3.b.2 ; it was just (2') and not (2) what we used there. Finally, $\overline{\mathcal{G}}_{\hat{\nu},\hat{\gamma}}$ is clearly a $\Sigma_{\Lambda}^{\omega}$-formula in view of lemma 3.b.1. Thus we can replace in theorem 3.b.3 $\mathcal{G}_{\hat{\nu},\hat{\gamma}}$ by $\overline{\mathcal{G}}_{\hat{\nu},\hat{\gamma}}$. – It is easily seen that $\overline{\mathcal{G}}_{\hat{\nu},\hat{\gamma}}$ does not imply $\mathcal{G}_{\hat{\nu},\hat{\gamma}}$. Consider e.g. the $\Sigma_{\Lambda}^{\omega}$-formula

$$\int(A) \equiv_{df} (\exists P).P(o) \wedge (\forall t)\lceil P(t') <\rightarrow \lceil P(t) <\rightarrow A(t)\rceil\rceil \wedge (\exists^{\omega}t)P(t) .$$

Then $\vartheta_{1,X,Y}(A,a,b)$ means "A contains an even (odd) number of F's between a and b" if X and Y are equal (different). Then there are 4 non-equivalent and satisfiable formulae $\mathfrak{H}_{\acute{\gamma}}$, and $\mathfrak{H}_4(A,a,b)$ has the meaning "A contains just one F between a and b". Clearly for any $\acute{\nu}$, $\mathfrak{G}_{\acute{\nu},4}$ is not satisfiable, whereas $\overline{\mathfrak{G}}_{\acute{\nu},4}(A)$ is equivalent to $(\exists^{\omega}t)\neg A(t)$. This example also shows that the set M is not independent from whether $\mathfrak{G}_{\acute{\nu},\acute{\gamma}}$ or $\overline{\mathfrak{G}}_{\acute{\nu},\acute{\gamma}}$ is used to construct it: for any $\acute{\nu}$, $(\acute{\nu},4)$ is in M in case $\mathfrak{G}_{\acute{\nu},\acute{\gamma}}$ is used, but not so in case of $\overline{\mathfrak{G}}_{\acute{\nu},\acute{\gamma}}$. Generally, the new set M is included in the old one; and if this inclusion is proper - far better for the construction.

There is only one hitch somewhere. Whereas we did not use the full meaning of the formulae $\mathfrak{G}_{\acute{\nu},\acute{\gamma}}$ in the proof of **lemma 3.b.2**, we do use it in the formal derivation of this lemma in 5.b. The proof in 5.b breaks down if (2) is weakened to (2'). Therefore we carried through the whole construction with $\mathfrak{G}_{\acute{\nu},\acute{\gamma}}$ instead of $\overline{\mathfrak{G}}_{\acute{\nu},\acute{\gamma}}$, since we want to prove the completeness of SC. But for the DP itself, of course we can use $\overline{\mathfrak{G}}_{\acute{\nu},\acute{\gamma}}$ instead of $\mathfrak{G}_{\acute{\nu},\acute{\gamma}}$. (For a final solution of the problem see 5.c.)

Therefore we can make the following changes in section b: Replace step (3) of part 3 by

(3') Construct formulae $\overline{\mathfrak{G}}_{\acute{\nu},\acute{\gamma}}$, $\acute{\nu},\acute{\gamma} = 1,\ldots,\ell$ as in section c.

Replace step (4) by

(4') Transform the $\overline{\mathfrak{G}}_{\acute{\nu},\acute{\gamma}}$ into Σ_1^{ω}-formulae using lemma 3.b.1. Delete part 5. Replace $\mathfrak{G}_{\acute{\nu},\acute{\gamma}}$ by $\overline{\mathfrak{G}}_{\acute{\nu},\acute{\gamma}}$ anywhere. At last change back the just introduced step (2') into the old one (2).

The replacing of $\mathfrak{G}_{\acute{\nu},\acute{\gamma}}$ by $\overline{\mathfrak{G}}_{\acute{\nu},\acute{\gamma}}$ gives a very considerable improvement of the DP what regards the growing of the length of the formulae: In the estimation of part 3 we can replace γ by $\overline{\gamma} =_{df} 2\gamma+3$, thus getting at the end

$$L(\neg \mathfrak{H}_{\acute{\gamma}}) < 2^{2^{2(\widetilde{m}_{\gamma}+1)}}(6\cdot2^{3\widetilde{m}_{\acute{\nu}}}+4) \ .$$

Especially we can replace now the terrible number at the end of section c by $2^{16}\cdot52-1 \approx 2^{21} \approx 10^6$ (which is of course not a small one, too).

Thus the most cumbersome part of the DP is now the number ℓ^2 of formulae $\overline{\mathfrak{G}}_{\acute{\nu},\acute{\gamma}}$, which we have already diminished considerably by the former considerations.

In most cases, the changes in part 3 and 5 will spare a lot of predicate quantifiers. Nevertheless, it is questionable whether it is possible to program the DP for wholly arbitrary formulae ever on a real computer. The most hopeful thing for people interested in practical application of the procedure (e.g. for analyzing processes, especially sequential circuits) will be to find simple types of formulae in which they could present their problems to the machine. The simplest such type would be Σ_1^ω itself, the next one Σ_2^ω with one universal and a few existential predicate quantifiers.

By the way, it is the thinking in processes which suggests a further improvement. Clearly, for the analysis of the course of a process the length of the unit time interval is irrelevant. Thus one can add in the changings to part 1: Perform step (2) before step (1). Let m be the greatest common divisor of all natural numbers which appear in terms of the formulae. Divide all these numbers by m, and then perform step (1) using lemmata 1 and 2 of this section. If $m > 1$, this changing will save a lot of new predicate quantifiers. - Further considerable improvements of the DP are to be found in 5.c and at the end of II.2.b.

d) Examples

Consideration of examples, though superflous, is sometimes helpful in understanding. Therefore we will perform the advices given in section a) and c) on two actually given formulae, simply stating the main results of each step.

Example 1: We start with the axiom (A2) written as a sentence:

$$(\forall x)\ x' \neq o$$

Part 1:   (2) $(\forall x)(\exists P).P(x') \wedge \neg P(o)$

(4) $(\forall S)(\exists P)(\exists x)(\forall z).S(z) \rightarrow P(x') \wedge \neg P(o) \wedge S(x)$

(5) $(\forall S)(\exists P).(\forall z)\neg S(z) \vee \ulcorner \neg P(o) \wedge (\exists x)\ulcorner S(x) \wedge P(x')\urcorner\urcorner$

(7)a) $(\forall S)(\exists P)(\exists R).\ulcorner R(o) \vee \ulcorner \neg R(o) \wedge \neg P(o)\urcorner\urcorner \wedge$

$\wedge\ (\forall t)\overline{\ulcorner}\ulcorner R(t) <\!\!-\!\!> R(t')\urcorner \wedge \overline{\ulcorner}\ulcorner R(t) \wedge \neg S(t)\urcorner \vee \neg R(t)\urcorner\overline{\urcorner}\wedge$

$\wedge\ (\exists t)\ \ulcorner R(t) \vee \ulcorner \neg R(t) \wedge S(t) \wedge P(t')\urcorner\urcorner$

   b) $(\forall S)(\exists P)(\exists R).\ulcorner R(o) \vee \neg P(o)\urcorner \wedge (\forall t)\overline{\ulcorner}\ulcorner R(t) <\!\!-\!\!> R(t')\urcorner \wedge$

$\wedge\ \ulcorner \neg S(t) \vee \neg R(t)\urcorner\overline{\urcorner} \wedge (\exists t)\ \ulcorner R(t) \vee \ulcorner S(t) \wedge P(t')\urcorner\urcorner$

Part 2:   (1) $\neg(\exists S)\neg(\exists P)(\exists R)(\exists Q).\neg Q(o) \wedge \ulcorner R(o) \vee \neg P(o)\urcorner \wedge$

$\wedge\ (\forall t)\overline{\ulcorner}\ulcorner R(t) <\!\!-\!\!> R(t')\urcorner \wedge \ulcorner \neg S(t) \vee \neg R(t)\urcorner \wedge$

$\wedge\ \ulcorner Q(t') <\!\!-\!\!> Q(t) \vee R(t) \vee \ulcorner S(t) \wedge P(t')\urcorner\overline{\urcorner}\overline{\urcorner} \wedge (\exists^{\omega} t)Q(t)$

(2) $(\exists P)(\exists R)(\exists Q).\neg Q(o) \wedge \ulcorner R(o) \vee \neg P(o)\urcorner \wedge (\forall t)\overline{\ulcorner}\ulcorner R(t) <\!\!-\!\!> R(t')\urcorner \wedge$

$\wedge\ \ulcorner \neg A(t) \vee \neg R(t)\urcorner \wedge \ulcorner Q(t') <\!\!-\!\!> Q(t) \vee R(t) \vee \ulcorner A(t) \wedge P(t')\urcorner\overline{\urcorner}\overline{\urcorner} \wedge$

$\wedge\ (\exists^{\omega} t)Q(t)$

Part 3: Since $m = 3$, we have to construct 64 formulae $\vartheta_{1,Y,Z}$ and 64 formulae $\vartheta_{2,Y,Z}$. It is easily seen, that we can cancel all pairs X,Z of the form

$$\begin{pmatrix} o \\ T \\ o \end{pmatrix},\ \begin{pmatrix} o \\ F \\ o \end{pmatrix}\quad\text{or}\quad \begin{pmatrix} o \\ F \\ o \end{pmatrix},\ \begin{pmatrix} o \\ T \\ o \end{pmatrix}\quad\text{or}\quad \begin{pmatrix} o \\ o \\ T \end{pmatrix},\ \begin{pmatrix} o \\ o \\ F \end{pmatrix},$$

where o stands for T or F, since those formulae are not satisfiable; from the 128 formulae rest 48. Further we cancel the pairs

$$\begin{pmatrix} o \\ F \\ T \end{pmatrix},\ \begin{pmatrix} o \\ F \\ T \end{pmatrix},$$

since the resulting formulae are valid. But the number of 44 remaining formulae is still too big to go on further. Thus we will start at this point with a new formula.

Remark that the above example deviates in some points from the official procedure: Clearly a real computer had to construct all the 128 formulae and then to test them for validity and satisfiability. Also, as it stands, step (7b) in part 1 is not in accord with the DP. Most programs for DPs have to be combined with a subprogram which simplifies formulae at suitable points by means of the propositional calculus. At last we have skipped (6), and incorporated its transformation into step (1) of part 2; the original version gives a much more complicated propositional structure of the formula. It will be often advisable to shift step (6) thus between (7) and (8), since then one may save predicate quantifiers in (8). In general, to save time it will be good to treat exceptional cases by extra advices.

__Example 2:__ We look for a $\Sigma_\wedge^\omega$-representation of the Σ_2^ω-sentence

$$(\forall P)(\exists R).\neg R(o) \wedge (\forall t)\lceil R(t') <\!-\!> R(t) \vee P(t)\rceil \wedge (\exists^\omega t)R(t)$$

Thus we define

$$\psi_2 \equiv_{df} (\exists R).\neg R(o) \wedge (\forall t)\lceil R(t') <\!-\!> R(t) \vee A(t)\rceil \wedge (\exists^\omega t)R(t)$$

__Part 3:__ (1) There are only two formulae ψ_γ which we have to take into consideration

$$\psi_{1,F,T} \equiv (\exists R).\neg R(a) \wedge (\forall t)\lceil R(t') <\!-\!> R(t) \vee A(t)\rceil \wedge R(b)$$

$$\psi_{1,F,F} \equiv (\exists R).\neg R(a) \wedge (\forall t)\lceil R(t') <\!-\!> R(t) \vee A(t)\rceil \wedge \neg R(b)$$

Among the remaining formulae are $\psi_{1,T,T}$ and $\psi_{2,T,T}$ valid, $\psi_{1,T,F}$ and $\psi_{2,T,F}$ and $\psi_{2,F,F}$ are not satisfiable, $\psi_{2,F,T}$ is equivalent to $\psi_{1,F,T}$.

(2) $\psi_{1,F,T} <\!-\!> \neg \psi_{1,F,F}$. Thus we may define $\gamma_1 \equiv_{df} \psi_{1,F,T}$, $\gamma_2 \equiv_{df} \psi_{1,F,F}$. Before going further it is good to ask for the meaning of these formulae. $\gamma_1(A,a,b)$ and $\gamma_2(A,a,b)$ are equivalent to $(\exists t)_a^b A(t)$ and $(\forall t)_a^b \neg A(t)$, resp. Therefore the formulae $\mathcal{G}_{i,j}$ are equivalent to the following:

$$\mathcal{G}_{1,1}(A) <\!-\!> (\exists^\omega t)A(t), \quad \mathcal{G}_{1,2}(A) <\!-\!> (\exists t)A(t) \wedge (\forall^\omega t)\neg A(t),$$

$$\mathcal{G}_{2,1}(A) <\!-\!> (\exists x)(\forall t)_o^x \neg A(t) \wedge (\exists^\omega t)A(t), \quad \mathcal{G}_{2,2}(A) <\!-\!> (\forall t)\neg A(t) .$$

This is what we meant in 2.c by "different types of ultimately periodic predicates": Since $\int_2(A)$ is equivalent to $(\exists t)A(t)$, it suffices to distinguish between good words (having a T) and bad ones (having no T). Thus there are four types of ultimately periodic predicates: both ini-

tial part and periodic germ good ($\mathcal{G}_{1,1}$) resp. bad ($\mathcal{G}_{2,2}$) and the
mixed types $\mathcal{G}_{1,2}$ and $\mathcal{G}_{2,1}$. It is clear from this discussion that
$\neg f_2 \iff \mathcal{G}_{2,2}$, thus $M = \{(2,2)\}$. We will not show this by the DP itself,
since it would be too expansive. But we will now apply part 5 to $\mathcal{G}_{2,2}$
to get a $\Sigma_\wedge^\omega$-representation for $\neg f_2$. This is clearly superflous in view
of section c, but perhaps helpful for grasping the proof of **theorem** 3.**a**.3.

Part 5: We consider the formula

$$(\exists Q).(\exists^\omega t)Q(t) \wedge (\exists x)\lceil Q(x) \wedge (\exists P)\lceil \neg P(o) \wedge$$

$$\wedge (\forall t)_o^x \lceil P(t') \iff P(t) \vee A(t)\rceil \wedge \neg P(x)\rceil\rceil \wedge$$

$$\wedge (\forall y)(\forall x)_o^y \lceil Q(x) \wedge Q(y) \to \mathcal{h}_2(A,x,y)\rceil .$$

$\mathcal{h}_1$ and $\mathcal{h}_2$ are already Σ_R^o-formulae; thus we skip step (1) and get by

(2) $(\exists P).\neg P(o) \wedge (\forall t)\lceil P(t') \iff P(t) \vee A(t)\rceil \wedge (\exists x)\lceil \neg P(x) \wedge Q(x)\rceil$

(3) $(\exists SR).S(o) \wedge (\forall t)_o^y \lceil \lceil S(t') \to S(t)\rceil \wedge \lceil S(t) \wedge \neg S(t') \to \neg R(t) \wedge Q(t)\rceil \wedge$

$\wedge \lceil \neg S(t') \to \lceil R(t') \iff R(t) \vee A(t)\rceil\rceil\rceil \wedge \neg S(y) \wedge R(y)$

(4) $(\exists \underline{R}^4).R_1(o) \wedge R_2(o) \wedge \neg R_3(o) \wedge \neg R_4(o) \wedge$

$\wedge (\forall t)_o^y \lceil \lceil R_1(t') \iff R_1(t) \vee R_3(t)\rceil \wedge$

$\wedge \lceil R_2(t') \iff R_2(t) \vee \lceil R_3(t) \wedge Q(t) \wedge A(t)\rceil \vee \lceil R_4(t) \wedge A(t)\rceil\rceil \wedge$

$\wedge \lceil R_3(t') \iff R_3(t) \vee R_1(t)\rceil \wedge$

$\wedge \lceil R_4(t') \iff \lceil R_3(t) \wedge Q(t) \wedge \neg A(t)\rceil \vee \lceil R_4(t) \wedge \neg A(t)\rceil\rceil\rceil \wedge R_2(y)$

(5) $(\exists \underline{R}^4).R_1(o) \wedge \ldots \wedge \neg R_4(o) \wedge (\forall t)_o^y \lceil \lceil \ldots \rceil \wedge \lceil \neg R_2(t) \vee \neg Q(t)\rceil \rceil$

(6) $(\exists PR^4 QS).\neg P(o) \wedge R_1(o) \wedge R_2(o) \wedge \neg R_3(o) \wedge \neg R_4(o) \wedge \neg S(o) \wedge$

$\wedge (\forall t)\lceil \lceil P(t') \iff P(t) \vee A(t)\rceil \wedge \lceil R_1(t') \iff R_1(t) \vee R_3(t)\rceil \wedge$

$\wedge \lceil R_2(t') \iff R_2(t) \vee \lceil R_3(t) \wedge Q(t) \wedge A(t)\rceil \vee \lceil R_4(t) \wedge A(t)\rceil\rceil \wedge$

$\wedge \lceil R_3(t') \iff R_3(t) \vee R_1(t)\rceil \wedge \lceil R_4(t') \iff \lceil R_3(t) \wedge Q(t) \wedge \neg A(t)\rceil \vee$

$\vee \lceil R_4(t) \wedge \neg A(t)\rceil\rceil \wedge \lceil \neg R_2(t) \vee \neg Q(t)\rceil \wedge$

$\wedge \lceil S(t') \iff S(t) \vee \lceil Q(t) \wedge \neg P(t)\rceil\rceil\rceil \wedge (\exists^\omega t)\lceil S(t) \wedge Q(t)\rceil$

This example shows that in most cases the upper bound of section b
is much too big: The formula we started with is of type $\forall\exists$ as considered
at the end of section b, but we got for f_2 an $\Sigma_\wedge^\omega$-representation of 7
predicate quantifiers instead of $10^{3 \cdot 10^6}$. Using $\overline{\mathcal{G}}_{2,2}$ instead of $\mathcal{G}_{2,2}$
we get a $\Sigma_\wedge^\omega$-formula of length 5, also much less than the number we got
at the end of section c. These astonishing improvements rest mostly on
the inserted steps (1a+b) and (2a+b) of section c.

For further exercise the reader may try to prove what was maintained in section c about the there defined formula f, and then to construct a $\Sigma_\wedge^\omega$-representation of $\neg f$.

§5. Derivations

a) The completeness of SC

At several places we have claimed that our axiom system of SC is complete, i.e. that for any sentence $\mathcal{O}$ holds: either $\mathcal{O}$ is derivable, or $\neg\,\mathcal{O}$ is. The DP presents an effective procedure which, applied to any sentence $\mathcal{O}$, yields a truth value X such that $\mathcal{O}$ is equivalent to X. Now we want to strengthen this to: There is an effective procedure which, applied to any sentence $\mathcal{O}$, yields a truth value X such that $\mathcal{O} <\!\!-\!\!> X$ is derivable in SC (for short $\vdash_{SC} \lceil \mathcal{O} <\!\!-\!\!> X \rceil$). Consequently $\vdash_{SC} \mathcal{O}$ or $\vdash_{SC} \neg\mathcal{O}$, depending on whether X is T or F, resp. To this end we have to derive in SC the assertions of the theorems and lemmata which justify the particular transformations of the DP. Then we have shown indeed the effective completeness of SC: there is an effective procedure which, applied to any sentence $\mathcal{O}$, yields a derivation in SC either for $\mathcal{O}$ or $\neg\,\mathcal{O}$.

To get a survey on the proof of the decidability we represent the course of the proof by a diagram. The numbers refer to theorems, lemmata (L), and corollaries (C), the arrows indicate the logical dependence. The remarks of 1.a are not mentioned in the diagram. Similarly, the recursion theorem 1.b.1 is connected in the diagram with the Ramsey theorem only, although it is used in most proofs.

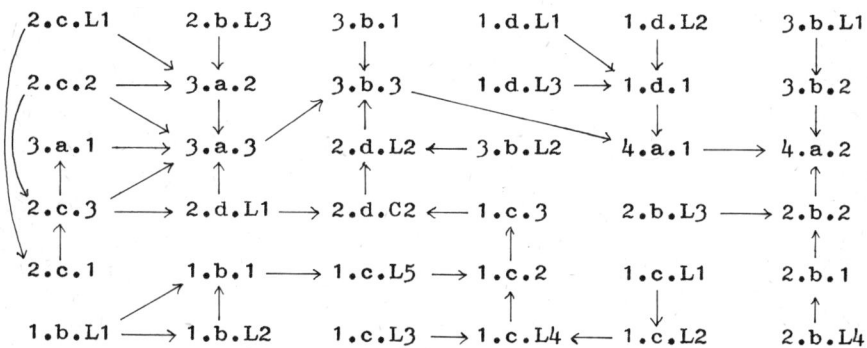

The proofs of 1.b and 1.c (the recursion theorem and the theorem of Ramsey) are given in such a manner that it is trivial (though time-consuming) to translate them into derivations of SC. From the recursion theorem follow easily the remarks 1)-7) in 1.a; all proofs concerning switching predicates use induction: one has to find out by the minimum principle the switching point.

To start with an easy example we will derive lemma 1 in 2.c:
Derivation of lemma 2.c.1: ->: Wholly analogous to lemma 1 in 1.b we
have

$$\lceil \underline{B}(a) <-> z \rceil \wedge (\forall t)_a^b \lceil \underline{B}(t') <-> \pounds[\underline{A}(t),\underline{B}(t)] \rceil \wedge \lceil \underline{C}(a) <-> z \rceil \wedge$$
$$\wedge (\forall t)_a^b \lceil \underline{C}(t') <-> \pounds[\underline{A}(t),\underline{C}(t)] \rceil -> (\forall t)_a^b \lceil \underline{B}(t) <-> \underline{C}(t) \rceil .$$

Further holds

$$\underline{B}(b) <-> \underline{C}(b) .->. \pounds[\underline{B}(b)] <-> \pounds[\underline{C}(b)] .$$

This gives together

$$\lceil \underline{B}(a) <-> z \rceil \wedge (\forall t)_a^b \lceil \underline{B}(t') <-> \pounds[\underline{A}(t),\underline{B}(t)] \wedge \pounds[\underline{B}(b)] .->.$$
$$.->. \lceil \underline{C}(a) <-> z \rceil \wedge (\forall t)_a^b \lceil \underline{C}(t') <-> \pounds[\underline{A}(t),\underline{C}(t)] \rceil -> \pounds[\underline{C}(b)] .$$

Quantification yields the wanted formula

$$(\exists \underline{P})\lceil \lceil \underline{P}(a) <-> z \rceil \wedge (\forall t)_a^b \lceil \underline{P}(t') <-> \pounds[\underline{A}(t),\underline{P}(t)] \rceil \wedge \pounds[\underline{P}(b)] \rceil ->$$
$$-> (\forall \underline{P})\lceil \lceil \underline{P}(a) <-> z \rceil \wedge (\forall t)_a^b \lceil \underline{P}(t') <-> \pounds[\underline{A}(t),\underline{P}(t)] \rceil -> \pounds[\underline{P}(b)] \rceil$$
$$\underline{<-}: (\forall \underline{P}) \lceil \vartheta(\underline{P}) -> \ell(\underline{P}) \rceil \wedge (\exists \underline{P}) \vartheta(\underline{P}) -> (\exists \underline{P}) \lceil \vartheta(\underline{P}) \wedge \ell(\underline{P}) \rceil$$

is derivable already in the logical frame. Choosing

$$\lceil \underline{P}(a) <-> z \rceil \wedge (\forall t)_a^b \lceil \underline{P}(t') <-> \pounds[\underline{A}(t),\underline{P}(t)] \rceil \quad \text{as} \quad \vartheta(\underline{P}) \quad \text{and}$$
$$\pounds[\underline{P}(b)] \quad \text{as} \quad \ell(\underline{P}) \quad ,$$

the second premise $(\exists \underline{P}) \vartheta(\underline{P})$ becomes an instance of the recursion theorem 1.b.1, and may thus be cancelled. The remaining formula is just what
we wanted to prove.

The derivation of lemma 2.b.3 is similar; theorem 2.c.1 follows directly from lemma 2.c.1. - To prove theorem 2.c.2 we need the following
lemma: Let the notations be as in the proof of theorem 2.c.2, let $\int^R$ be
the Σ_R^o -formula constructed there from $\int$. Define Σ^o -formulae $\int_\ell$,
$\ell = 1,\ldots,m$ by replacing the output condition $\pounds[\underline{P}(b)]$ in $\int$ by
$\lceil \underline{P}(b) <-> Y_\ell \rceil$, analogously define Σ_R^o -formulae $\int_\ell^R$, $\ell = 1,\ldots,m$ from $\int^R$
with the output condition $R_\ell(b)$.

Lemma 1: $a \leq b -> \bigwedge_{\ell=1}^m \lceil \int_\ell(a,b) <-> \int_\ell^R(a,b) \rceil$ is derivable in SC.

Proof: Induction over b:
a) a = b : For any ℓ, $\int_\ell(a,a)$ is equivalent to

$$(\exists \underline{P}^*). \alpha[\underline{P}(a)] \wedge \lceil \underline{P}(a) <-> Y_\ell \rceil ,$$

$\int_\ell^R(a,a)$ is equivalent to

$$(\exists \underline{R}^m). \bigwedge_{j=1}^m \lceil R_j(a) <-> \alpha[Y_j] \rceil \wedge R_\ell(a) .$$

Both formulae are with the help of (COMP) equivalent to $\alpha[Y_\ell]$.

b)   Induction step: Let the lemma be proved for $b \geq a$. Let ℓ be fixed,
suppose first $f_\ell(a,b')$. Thus we have

$$(\exists \underline{P}). \, \mathcal{O}[\underline{P}(a)] \wedge (\forall t)_a^b \mathcal{L}[\underline{A}(t),\underline{P}(t),\underline{P}(t')] \wedge \overset{m}{\underset{\nu=1}{\bigvee}} \ulcorner \underline{P}(b) \mathrel{<\!\!-\!\!>} Y_\nu \urcorner \wedge$$

$$\wedge \, \mathcal{L}[\underline{A}(b),\underline{P}(b),\underline{P}(b')] \wedge \ulcorner \underline{P}(b') \mathrel{<\!\!-\!\!>} Y_\ell \urcorner$$

Let $\tilde{f}_\nu(\underline{A},\underline{B}^{\kappa},a,b)$ be the kernel of $f_\nu(\underline{A},a,b)$ (i.e. $f_\nu(\underline{A},a,b) \equiv$
$\equiv (\exists \underline{P}) \tilde{f}_\nu(\underline{A},\underline{P},a,b))$, analogously let $\tilde{f}_\nu^R(\underline{A},\underline{B},a,b)$ be the kernel of
$f_\nu^R(\underline{A},a,b)$, $\nu = 1,\ldots,m$. Then we have shown: There is a γ such that

$$(\exists \underline{P}^{\kappa}). \, \tilde{f}_\gamma(\underline{A},\underline{P},a,b) \quad \wedge \, \mathcal{L}[\underline{A}(b),\underline{P}(b),\underline{P}(b')] \wedge \ulcorner \underline{P}(b') \mathrel{<\!\!-\!\!>} Y_\ell \urcorner$$

From this follows

(1) $\mathcal{L}[\underline{A}(b),Y_\gamma,Y_\ell]$

and further by the induction hypothesis

(2) $f_\gamma^R(\underline{A},a,b)$.

Using the recursion theorem in the form

$$(\exists \underline{R}^m). \ulcorner \underline{R}(a) \mathrel{<\!\!-\!\!>} \mathcal{U}^m \urcorner \wedge (\forall t)_a^{b'} \ulcorner \underline{R}(t') \mathrel{<\!\!-\!\!>} \mathcal{L}^m(t) \urcorner$$

we get $f_\ell^R(\underline{A},a,b')$ from (1) and (2). Analogously we prove

$$f_\ell^R(\underline{A},a,b') \rightarrow f_\ell(\underline{A},a,b') .$$

The lemma follows. #

   From the lemma we conclude

$$\underset{\ell \in M}{\bigwedge} \ulcorner f_\ell(\underline{A},a,b) \mathrel{<\!\!-\!\!>} f_\ell^R(\underline{A},a,b) \urcorner ,$$

from which we get theorem 2.c.2 by a simple transformation.

   The theorems 2.c.3 and 3.a.1+2 follow directly from theorem 2.c.2.
Since lemma 2.d.1 is trivial, we have further theorem 3.a.3. The proof
of lemma 3.b.2 is to be found in section b. Lemma 2.d.2 and the remaining parts of 3.b (lemma 1, theorems 1-3) are easy; so are the lemmata
and the theorem of 1.d. These derivations give together the derivability
of the normal form theorem 4.a.1: To any formula $f(\underline{A})$ one can construct
effectively a formula $g(\underline{A})$ in Σ_1^ω such that $f(\underline{A}) \mathrel{<\!\!-\!\!>} g(\underline{A})$ is derivable
in SC.

   It remains to derive lemma 4 and theorem 1 in 2.b, and from this
theorem 2.b.2. These proofs are a little more cumbersome, since we have
to use the fact, not only that 0_m is finite, but also that there are
only finitely many m-words of a certain length.

   To begin with lemma 2.b.4 let $g(a,b)$ be a sentence in Σ^0 (i.e. a
Σ^0-formula without free predicate variables, which contains of course
the individual variables a and b),

$$\mathcal{G}(a,b) \equiv_{df} (\exists \underline{P}^{m}). \, \alpha[\underline{P}(a)] \wedge (\forall t)_a^b \mathcal{L}[\underline{P}(t),\underline{P}(t')] \wedge \mathcal{L}[\underline{P}(b)],$$

let $\mathcal{R} =_{df} 2^m$:

Lemma 2: $(\exists x) \mathcal{G}(o,x) <\!-\!> (\exists x)_o^{\mathcal{R}} \mathcal{G}(o,x)$ is derivable in SC.

Proof: Suppose $(\exists x) \mathcal{G}(o,x)$, let b be minimal such that $\mathcal{G}(o,b)$. Assume: $b \geq \mathcal{R}$. Let $\underline{B}$ be a carrying predicate for $\mathcal{G}(o,b)$, thus

$$\alpha[\underline{B}(o)] \wedge (\forall t)_a^b \mathcal{L}[\underline{B}(t),\underline{B}(t')] \wedge \mathcal{L}[\underline{B}(b)] \quad .$$

We split up this into

(1) $\quad \alpha[\underline{B}(o)] \wedge \bigwedge_{\nu=0}^{\mathcal{R}-1} \mathcal{L}[\underline{B}(\nu),\underline{B}(\nu+1)] \wedge (\forall t)_{\mathcal{R}}^b \mathcal{L}[\underline{B}(t),\underline{B}(t')] \wedge \mathcal{L}[\underline{B}(b)].$

Surely holds

$$\bigwedge_{\jmath=1}^{\mathcal{R}} \bigvee_{\nu=1}^{\mathcal{R}} \lceil \underline{B}(\jmath) <\!-\!> Y_\nu \rceil \quad .$$

Passing to the disjunctive normal form yields

$$\bigvee_{\nu_1,\dots,\nu_{\mathcal{R}}=1}^{\mathcal{R}} \bigwedge_{\jmath=0}^{\mathcal{R}} \lceil \underline{B}(\jmath) <\!-\!> Y_{\nu_\jmath} \rceil \quad ,$$

from which we conclude

$$\bigvee_{0 \leq \ell < \jmath \leq \mathcal{R}} \lceil \underline{B}(\ell) <\!-\!> \underline{B}(\jmath) \rceil \quad .$$

Let $\ell < \jmath$ be fixed with this property. Then we get from (1)

(2) $\quad \alpha[\underline{B}(o)] \wedge \bigwedge_{\nu=0}^{\ell-1} \mathcal{L}[\underline{B}(\nu),\underline{B}(\nu+1)] \wedge \lceil \underline{B}(\ell) <\!-\!> \underline{B}(\jmath) \rceil \wedge$

$\qquad \wedge \bigwedge_{\nu=\jmath}^{\mathcal{R}-1} \mathcal{L}[\underline{B}(\nu),\underline{B}(\nu+1)] \wedge (\forall t)_{\mathcal{R}}^b \mathcal{L}[\underline{B}(t),\underline{B}(t')] \wedge \mathcal{L}[\underline{B}(b)]$

Define $\underline{C}^m$ by

$$\underline{C}(c) \equiv_{df} \begin{cases} \underline{B}(a); & c < \ell \\ \underline{B}(a+\jmath-\ell); & \ell \leq c \end{cases}$$

Then (2) turns into

$$\alpha[\underline{C}(o)] \wedge \bigwedge_{\nu=0}^{\mathcal{R}-\jmath+\ell-1} \mathcal{L}[\underline{C}(\nu),\underline{C}(\nu+1)] \wedge (\forall t)_{\mathcal{R}-\jmath+\ell}^{b-\jmath+\ell} \mathcal{L}[\underline{C}(t),\underline{C}(t')] \wedge$$
$$\wedge \mathcal{L}[\underline{C}(b)] \quad ,$$

from which follows

$$\mathcal{G}(o,b-\jmath+\ell), \quad \text{what is a contradiction.}$$

To indicate the formalization of this proof we shorten the formula (2) by $\mathcal{G}_{\ell,\jmath}(\underline{B})$. Then we have proved (where $\tilde{\mathcal{G}}$ is the kernel of $\mathcal{G}$)

(3) $\quad \tilde{\mathcal{G}}(\underline{B},o,b) \wedge \mathcal{R} \leq b \, -\!> \bigvee_{0 \leq \ell < \jmath \leq \mathcal{R}} \mathcal{G}_{\ell,\jmath}(\underline{B}) \quad ,$

and further for any $\ell < \jmath$

$$\mathcal{G}_{\ell,\jmath}(\underline{B}) \wedge \bigwedge_{\nu=1}^{m} \lceil (\forall t) \lceil C_\nu(t) <\!-\!> \lceil t < \ell \wedge B_\nu(t) \rceil \vee$$
$$\vee \lceil \ell \leq t \wedge B_\nu(t+\jmath-\ell) \rceil \rceil \rceil \rceil \, -\!> \tilde{\mathcal{G}}(\underline{C},o,b-\jmath+\ell) \quad .$$

Quantification gives

(4) $(\exists \underline{P}) \lceil g_{\ell,j}(\underline{P}) \wedge (\exists \underline{R})(\forall t) \lceil \underline{R}(t) <-> \underline{\vartheta}(\underline{P},t) \rceil \; -> (\exists \underline{R}) \tilde{g}(\underline{R},o,b-j+\ell)$,

where we have shortened the defining formula for $C_{\nu}(t)$ by $\vartheta_{\nu}(\underline{B},t)$.
With the help of (COMP) we have

$(\exists \underline{R})(\forall t) \lceil \underline{R}(t) <-> \underline{\vartheta}(\underline{B},t) \rceil$.

Thus we may replace (3) by

$$\tilde{g}(\underline{B},o,b) \wedge \mathscr{R} \leq b -> \bigvee_{0 \leq \ell < \tilde{j} \leq \mathscr{R}} \lceil g_{\ell,j}(\underline{B}) \wedge$$
$$\wedge (\exists \underline{R}(\forall t) \lceil \underline{R}(t) <-> \vartheta(\underline{B},t) \rceil \rceil .$$

Quantification of $\underline{B}$ gives together with (4) the wanted result

$$g(o,b) \wedge \mathscr{R} \leq b -> \bigvee_{0 \leq \ell < \tilde{j} \leq \mathscr{R}} g(o,b-j+\ell). \quad \#$$

If we replace in $(\exists x)_o^{\mathscr{R}} g(o,x)$ the bound quantifier $(\exists x)_o^{\mathscr{R}}$ by a disjunction, we get the equivalent formula

$$\bigvee_{\nu=0}^{\mathscr{R}-1} g(o,\nu) .$$

If we replace further in $g(o,\nu)$ the bound quantifier $(\forall t)_o^{\nu}$ by a conjunction and the quantifier $(\exists \underline{P})$ by a disjunction over all words of length $\nu+1$, we get

$$\bigvee_{(X_o,\dots,X_\nu)} \lceil \alpha[X_o] \wedge \bigwedge_{\tilde{j}=0}^{\nu-1} \mathscr{L}[X_{\tilde{j}},X_{\tilde{j}+1}] \wedge \mathscr{L}[X_\nu] ,$$

which is a propositional formula and thus decidable. Thus we have shown:
There is an effective procedure which produces to any sentence $g(a,b)$
in Σ^o a derivation of the formula $(\exists x) g(o,x)$, or of its negation. This
statement is the formal counterpart to lemma 2.b.4, since the satis-
fiability of a Σ^o-formula $\mathfrak{f}(\underline{A}^w,a,b)$ is equivalent to the truth of the
sentence $(\exists x)(\exists \underline{R}^w) \mathfrak{f}(\underline{R},o,x)$ and $(\exists \underline{R}) \mathfrak{f}(\underline{R},a,b)$ is a Σ^o-sentence (in the
above sense). - Theorem 2.b.1 follows directly from lemma 2.b.4.

For the derivation of theorem 2.b.2 take over the notations from the
proof given in 2.b. Let $M = \{ Y \varepsilon O_m; \mathscr{L}[Y] \}$

Lemma 3: $g <-> \bigvee_{Y \varepsilon M} \lceil (\exists y)(\exists x)_{o,}^y \vartheta_Y(o,x) \wedge \ell_Y(x,y) \rceil$ is derivable in SC.

Proof: ->: Let $\underline{B}$ be a carrying predicate of g:

$$\alpha[\underline{B}(o)] \wedge (\forall t) \mathscr{L}[\underline{B}(t),\underline{B}(t')] \wedge (\exists^w t) \mathscr{L}[\underline{B}(t)] .$$

From the final condition follows the existence of numbers $o < a_1 < a_2 <$
$< \dots < a_{\mathscr{R}+1}$ such that $\underline{B}(a_\nu) \varepsilon M$, $\nu = 1,\dots,\mathscr{R}+1$.
From this we conclude as in the proof of lemma 2

$$(\exists y)(\exists x)_{o,}^y . \mathscr{L}[\underline{B}(x)] \wedge \mathscr{L}[\underline{B}(y)] \wedge \lceil \underline{B}(x) <-> \underline{B}(y) \rceil ,$$

and therefore

$$\bigvee_{Y \in M} \cdot (\exists y)(\exists x)_{o'}^{y} \lceil \alpha[\underline{B}(o)] \wedge (\forall t)_{o}^{x} \mathcal{L}\lfloor \underline{B}(t), \underline{B}(t') \rfloor \wedge \lceil \underline{B}(x) <-> Y \rceil \wedge$$

$$\wedge \ (\forall t)_{x}^{y} \mathcal{L}[\underline{B}(t), \underline{B}(t')] \wedge \lceil \underline{B}(y) <-> Y \rceil \rceil \ ,$$

which gives the desired conclusion.

<u><-:</u> Let Y be in M, let $\tilde{\vartheta}_Y$ and $\tilde{\ell}_Y$ be the kernel of ϑ_Y and ℓ_Y, resp.
Suppose

(1) $(\exists y)(\exists x)_{o'}^{y} \lceil \vartheta_Y(o,x) \wedge \ell_Y(x,y) \rceil$.

$(\exists y)_{x'} \ell_Y(x,y)$ is by lemma 2 equivalent to $(\exists y)_{x'}^{x+R} \ell_Y(x,y)$, and this
again is equivalent to $\bigvee_{\nu=1}^{R-1} \ell_Y(x,x+\nu)$. If we substitute this into (1),

we get by distributing:
There is a number $\hat{\gamma}$, $1 \leq \hat{\gamma} < R$, so that $(\exists x)_o \lceil \vartheta_Y(o,x) \wedge \ell_Y(x,x+\hat{\gamma}) \rceil$.
Let w be a number so that $\vartheta_Y(o,w) \wedge \ell_Y(w,w+\hat{\gamma})$, let $\underline{B}$ and $\underline{C}$ be carrying
predicates for $\vartheta_Y(o,)$ and $\ell_Y(w,w+\hat{\gamma})$, resp. Define

$$\underline{D}(a) \equiv_{df} \begin{cases} \underline{B}(a) \ ; \ a < w \\ \underline{C}(w+\nu) \ ; \ w \leq a \wedge a \equiv \nu \ (\hat{\gamma}) \end{cases}$$

Then we have

$$\alpha[\underline{D}(o)] \wedge (\forall t) \mathcal{L}[\underline{D}(t), \underline{D}(t')] \wedge (\forall t) \lceil t \equiv o(\hat{\gamma}) \ -> \ \mathcal{L}[\underline{D}(t)] \rceil \ ,$$

which gives $\mathcal{G}$.
The exact formal proof comes off in a way as indicated at the end of the
proof of lemma 2. #

   Lemmata 2 and 3 together give very easily theorem 2.b.2, which implies theorem 4.a.2. Thus we conclude:

<u>Theorem 1:</u> There is an effective procedure which applied to any sentence
$\mathcal{G}$ yields a derivation in SC for either $\mathcal{G}$ or $\neg \mathcal{G}$. In other words: The
axiom system for SC is complete.

b) Minimal Σ°-predicates

The last gap in the proof of the completeness of SC is the derivation of lemma 3.b.2. If we try to translate the proof of section 3.b into a SC-derivation, we fall into difficulties. We fix $i, \gamma, \underline{A}, \underline{B}$ as there and suppose $\mathcal{G}_{i,\gamma}(\underline{A}) \wedge \mathcal{G}_{i,\gamma}(\underline{B}) \wedge f(\underline{A})$. Then we have (corresponding to the sequences (a_γ) and (b_γ)) predicates E and H so that

(1) $(\exists^\omega t)E(t) \wedge \mathcal{h}_{\gamma'}(\underline{A},o,a) \wedge (\forall y)(\forall x)_0^y \lceil E(x) \wedge E(y) \to \mathcal{h}_{\gamma'}(\underline{A},x,y) \rceil$

(2) $(\exists^\omega t)H(t) \wedge \mathcal{h}_{\gamma'}(\underline{B},o,b) \wedge (\forall y)(\forall x)_0^y \lceil H(x) \wedge H(y) \to \mathcal{h}_{\gamma'}(\underline{B},x,y) \rceil$

There a and b are the first elements of E and H, resp. Let $\underline{C}$ be a carrying predicate for $f(\underline{A})$:

(3) $\mathcal{O}\!\lceil \underline{C}(o) \rceil \wedge (\forall t)\mathcal{J}\lceil \underline{A}(t),\underline{C}(t),\underline{C}(t') \rceil \wedge (\exists^\omega t)\mathcal{L}\lceil \underline{C}(t) \rceil$.

The first difficulty is that we cannot express within SC the one-to-one correspondence between the elements of E and H which is indicated by the indices of the sequences. We can show the existence of carrying words for $f(\underline{B})$ between any two elements of H, but we cannot construct them one after the other numbering them $\underline{D}_1(b_1)...\underline{D}_1(b_2)$, $\underline{D}_2(b_2)...$
$..\underline{D}_2(b_3),....$.

This difficulty is avoided by the following construction which yields a state Z^w so that the defined words are not only equivalent (lie in $T(\mathcal{h}_{\gamma'})$), but moreover the corresponding carrying words all start and end with the same state Z: Use

$$(\exists^\omega t)E(t) \wedge (\exists^\omega t)\mathcal{L}\lceil \underline{C}(t) \rceil$$

to construct a sequence of $2(f+1)$ numbers so that

$$a < a_0 \leq \tilde{a}_0 < a_1 \leq \tilde{a}_1 < ... < a_f \leq \tilde{a}_f \wedge \bigwedge_{\gamma=0}^{f} E(a_\gamma) \wedge \mathcal{L}\lceil \underline{C}(\tilde{a}_\gamma) \rceil .$$

Since O_w has only f elements, there must be a repetition among the states $\underline{C}(a_\gamma)$, i.e. there are numbers $o \leq f < g \leq f$ so that $\underline{C}(a_f) \equiv \underline{C}(a_g)$. Let $X \equiv_{df} \underline{C}(o)$, $Y \equiv_{df} \underline{C}(a)$, $Z \equiv_{df} \underline{C}(a_f)$, let $\tilde{b}$ be the second element of H. Then we have shown

$$\vartheta_{1,X,Y}(\underline{A},o,a) \wedge \vartheta_{1,Y,Z}(\underline{A},a,a_f) \wedge \vartheta_{2,Z,Z}(\underline{A},a_f,a_g) .$$

From this we get by methods as in the proof of lemma 3.b.2

$$\vartheta_{1,X,Y}(\underline{B},o,b) \wedge \vartheta_{1,Y,Z}(\underline{B},b,\tilde{b}) \wedge$$

$$\wedge (\forall y)(\forall x)_0^y \lceil H(x) \wedge H(y) \to \vartheta_{2,Z,Z}(\underline{B},x,y) \rceil .$$

Splicing the first two conjunctions together to $\vartheta_{1,X,Z}(\underline{B},o,\tilde{b})$ we seem to be ready: The carrying words of $\vartheta_{2,Z,Z}(\underline{B},x,y)$ shown to exist can be spliced together, since they all start and end with the same state Z.

But it is here where the next difficulty arises. The methods of proof

of lemmata 2 and 3 in section a are easily generalized to splice together any fixed number of suitable predicates. But we cannot carry over
this method to infinitely many predicates, since we cannot index them by
individual variables. The situation is analogous to the proof of the recursion theorem where the infinitely many initial parts of a recursively
introduced predicate are put together. Thus one might try to define as
there for $\gamma = 1, \ldots, m$

$$D_{\gamma}(d) \leftrightarrow \lceil d < \tilde{b} \wedge (\exists \underline{P}) \lceil \tilde{\vartheta}_{1, X, Z}(\underline{B}, \underline{P}, o, \tilde{b}) \wedge P_{\gamma}(d) \rceil \rceil \ \vee$$

$$\vee \ (\exists y)(\exists x)_{b}^{y} \lceil H(x) \wedge H(y) \wedge (\exists \underline{P}) \lceil \tilde{\vartheta}_{2, Z, Z}(\underline{B}, \underline{P}, x, y) \wedge P_{\gamma}(d) \rceil \rceil$$

where $\tilde{\vartheta}_{1}$, $\tilde{\vartheta}_{2}$ are the kernels of ϑ_{1}, ϑ_{2}, resp. The existence of $\underline{D}$ defined in this way follows from (COMP). But (in contrary to the situation
in section 1.b) the carrying predicates of ϑ_{1}, ϑ_{2} are not uniquely determined, what can effect that $\underline{D}$ does not even satisfy the transition
condition $\mathcal{G}[\underline{B}(d), \underline{D}(d), \underline{D}(d')]$. The only way-out is to force the carrying predicates of ϑ_{1}, ϑ_{2} to be uniquely determined. This will be done
by the following idea: The elements of 0_{m}, being finitely many, can be
ordered, e.g. lexicographically. This order can be extended in a natural
way to the words of a fixed length. The resulting order is definable in
SC by a formula having as a free variable the length of the words considered. Therefore we may speak of a carrying word minimal with respect
to a Σ^{o}-formula.

<u>Definitions:</u> 1) $Y, Z \ \epsilon \ 0_{m} : \underline{Y < Z} \ \equiv_{df} \ \bigvee_{\nu=1}^{m} . \bigwedge_{\gamma=1}^{\nu-1} \lceil Y_{\gamma} \leftrightarrow Z_{\gamma} \rceil \wedge Y_{\nu} \wedge \neg Z_{\nu}$

2) $\underline{A}, \underline{B} \ \epsilon \ S_{m} : \underline{A \overset{b}{\underset{a}{\equiv}} B} \ \equiv_{df} \ (\forall t)_{a}^{b} \lceil A(t) \leftrightarrow B(t) \rceil$

3) $\underline{A \overset{b}{\underset{a}{<}} B} \ \equiv_{df} \ (\exists x)_{a}^{b} . \underline{A}(x) < \underline{B}(x) \wedge \underline{A \overset{x}{\underset{a}{\equiv}} B}$

4) $\underline{A \overset{b}{\underset{a}{\leq}} B} \ \equiv_{df} \ \underline{A \overset{b}{\underset{a}{\equiv}} B} \vee \underline{A \overset{b}{\underset{a}{<}} B}$

The following remarks are easily derived in SC:

<u>Remarks:</u> 1) $\overset{b}{\underset{a}{<}}$ and $\overset{b}{\underset{a}{\leq}}$ are (total) ordering relations on predicates:

a) $\neg \underline{A \overset{b}{\underset{a}{<}} A}$

b) $\underline{A \overset{b}{\underset{a}{<}} B} \wedge \underline{B \overset{b}{\underset{a}{<}} C} \rightarrow \underline{A \overset{b}{\underset{a}{<}} C}$

c) $\underline{A \overset{b}{\underset{a}{<}} B} \vee \underline{B \overset{b}{\underset{a}{<}} A} \vee \underline{A \overset{b}{\underset{a}{\equiv}} B}$

d) $\underline{A \overset{b}{\underset{a}{\leq}} A}$

e) $\underline{A \overset{b}{\underset{a}{\leq}} B} \wedge \underline{B \overset{b}{\underset{a}{\leq}} C} \rightarrow \underline{A \overset{b}{\underset{a}{\leq}} C}$

f) $\underline{A} \overset{b}{\underset{a}{\lessgtr}} \underline{B} \lor \underline{B} \overset{b}{\underset{a}{\lessgtr}} \underline{A}$

g) $\underline{A} \overset{b}{\underset{a}{\lessgtr}} \underline{B} \land \underline{B} \overset{b}{\underset{a}{\lessgtr}} \underline{A} \rightarrow \underline{A} \overset{b}{\underset{a}{\equiv}} \underline{B}$

2) The strict ordering is permanent in time, the weak ordering only under suitable continuation:

a) $\underline{A} \overset{b}{\underset{a}{\leq}} \underline{B} \land b \leq c \rightarrow \underline{A} \overset{c}{\underset{a}{\leq}} \underline{B}$

b) $\underline{A} \overset{b}{\underset{a}{\leq}} \underline{B} \land b \leq c \land \underline{A} \overset{c}{\underset{b}{\leq}} \underline{B} \rightarrow \underline{A} \overset{c}{\underset{a}{\leq}} \underline{B}$

Let $f(\underline{A},a,b)$ be a Σ°-formula with kernel $\mathcal{G}$:

$(\exists \underline{P}^m)\, \mathcal{G}(\underline{A},\underline{P},a,b) \equiv (\exists \underline{P}^m)\lceil \mathcal{A}[\underline{P}(a)] \land (\forall t)_a^b \mathcal{L}[\underline{A}(t),\underline{P}(t),\underline{P}(t')] \land \mathcal{S}[\underline{P}(b)]\rceil$.

Definition 5: $\underline{B}$ is a <u>minimal carrying predicate</u> for $f(\underline{A},a,b)$, if it satisfies the condition

$\mathrm{Min}_{\mathcal{G}}(\underline{B};\underline{A},a,b) \equiv_{df} \mathcal{G}(\underline{A},\underline{B},a,b) \land (\forall \underline{R})\lceil \mathcal{G}(\underline{A},\underline{R},a,b) \rightarrow \underline{B} \overset{b'}{\underset{a}{\leq}} \underline{R}\rceil$.

Before we can prove the existence of minimal carrying predicates for f, we must show the uniqueness (lemma 1), and then the existence for the associated formulae f_ℓ (theorem 1).

Lemma 1: $\mathrm{Min}_{\mathcal{G}}(\underline{B};\underline{A},a,b) \land \mathrm{Min}_{\mathcal{G}}(\underline{C};\underline{A},a,b) \rightarrow \underline{B} \overset{b'}{\underset{a}{\equiv}} \underline{C}$ is derivable in SC. The proof is obvious.

Let $Y_1, Y_2, \ldots, Y_{\mathcal{R}}$ be the elements of O_m. Analogously to section a define for $\ell = 1, \ldots, \mathcal{R}$

$\mathcal{G}_\ell(\underline{A},\underline{B},a,b) \equiv_{df} \mathcal{A}\lfloor \underline{B}(a)\rfloor \land (\forall t)_a^b \mathcal{L}[\underline{A}(t),\underline{B}(t),\underline{B}(t')] \land \lceil \underline{B}(b) \longleftrightarrow Y_\ell \rceil$.

Theorem 1: $\bigwedge_{\ell=1}^{\mathcal{R}} \lceil (\exists \underline{P})\, \mathcal{G}_\ell(\underline{A},\underline{P},a,b) \rightarrow (\exists \underline{P})\, \mathrm{Min}_{\mathcal{G}_\ell}(\underline{P};\underline{A},a,b)\rceil$ is der. in SC.

Proof: a) Induction over b:

Trivial for b = a, since for any ℓ both the premise and the conclusion are equivalent to $(\exists \underline{P})\lceil \underline{P}(a) \longleftrightarrow Y_\ell \rceil$. So let the theorem be proved for b , suppose

(1) $(\exists \underline{P})\, \mathcal{G}_\ell(\underline{A},\underline{P},a,b')$

for a fixed ℓ. Let a and b be fixed for the moment (we shall indicate later on the translation into a formal proof), and define

$M =_{df} \{ \nu ;\ 1 \leq \nu \leq \mathcal{R} \land (\exists \underline{P}) \lceil \mathcal{G}_\nu(\underline{A},\underline{P},a,b) \land \mathcal{L}[\underline{A}(b),Y_\nu,Y_\ell]\rceil \}$.

By (1), $M \neq \emptyset$. With the help of M we look now for a minimal predicate among the carrying predicate of (1).

By induction hypothesis, predicates $\underline{D}_\nu$ exist for any $\nu \epsilon M$ such that

(2) $\mathrm{Min}_{\mathcal{G}_\nu}(\underline{D}_\nu;\underline{A},a,b)$.

For any pair $i \neq j$, where $i, j \in M$, there exists a number $a_{i,j}$ such that

$$a \leq a_{i,j} \leq b \wedge \underline{D}_i \overset{a_{i,j}}{\equiv} \underline{D}_j \wedge \neg \lceil \underline{D}_i(a_{i,j}) <-> \underline{D}_j(a_{i,j}) \rceil$$

(since $\neg \lceil \underline{D}_i(b) <-> \underline{D}_j(b) \rceil$). Clearly $a_{i,j} = a_{j,i}$. Let $a_{i,i} =_{df} a$, define
an ordering on M by

$$i <_M j \equiv_{df} \underline{D}_i(a_{i,j}) < \underline{D}_j(a_{i,j})$$

Let j be the smallest element under $<_M$: For any $i \neq j$ holds

$$\underline{D}_i \overset{a_{i,j}}{\underset{a}{\equiv}} \underline{D}_j \wedge \underline{D}_j(a_{i,j}) < \underline{D}_i(a_{i,j}) \ ,$$

which yields by definition $\underline{D}_j \overset{a_{i,j}}{\underset{a}{\leq}} \underline{D}_i$.
From remark 2a) follows

(3) $\underline{D}_j \overset{b'}{\underset{a}{\leq}} \underline{D}_i$ for every $i \in M$, $i \neq j$.

Change $\underline{D}_j$ at place b' into Y_ℓ. We want to show:

(4) $\text{Min}_{y_\ell}(\underline{D}_j; \underline{A}, a, b')$.

By (2) and the definition of M follows

$$\mathcal{G}_\ell(\underline{A}, \underline{D}_j, a, b') \ .$$

Suppose $\underline{C}$ so that $\mathcal{G}_\ell(\underline{A}, \underline{C}, a, b')$. Then $\underline{C}(b) \equiv Y_i$ for some $i \in M$.

1.case: $i = j$. Then $\underline{D}_j \overset{b'}{\underset{a}{\leq}} \underline{C}$ by (2). Since further $\underline{C}(b') \equiv \underline{D}(b') [\equiv Y_\ell]$,
we get from remark 2b)

$$\underline{D}_j \overset{b''}{\underset{a}{\leq}} \underline{C} \ .$$

2.case: $i \neq j$. Then $\underline{D}_j \overset{b'}{\underset{a}{\leq}} \underline{D}_i \leq \underline{C}$ by (3) and (2).

Remarks 1b) and 2a) give

$$\underline{D}_j \overset{b''}{\underset{a}{\leq}} \underline{C} \ .$$

Thus (4) follows in both cases.

b) Now we will give some details for the formalizing of the proof: Let
M be any subset of the set $\{1, \ldots, k\}$ so that $\bigwedge_{i \in M} \mathcal{G}[A(b), Y_i, Y_\ell]$ holds,
use $i <_M j$ as abbreviation for $\underline{D}_i(a_{i,j}) < \underline{D}_j(a_{i,j})$. Then we derive
easily within SC:

$$\bigwedge_{\substack{i,j \in M \\ i \neq j}} \lceil a \leq a_{i,j} \leq b \wedge \underline{D}_i \overset{a_{i,j}}{\underset{a}{\equiv}} \underline{D} \wedge \neg \lceil \underline{D}_i(a_{i,j}) <-> \underline{D}_j(a_{i,j}) \rceil \rceil \ ->$$

$$-> \bigwedge_{i \in M} \neg\, i <_M i \wedge \bigwedge_{\substack{i,j \in M \\ i \neq j}} \lceil i <_M j \vee j <_M i \rceil \wedge$$

$$\wedge \bigwedge_{i,j,j \in M} \lceil i <_M j \wedge j <_M j \ -> \ i <_M j \rceil \ .$$

From the conclusion (expressing that $<_M$ is a ordering on M) follows by means of the propositional calculus

$$\bigvee_{\mathcal{f} \in M} \bigwedge_{\substack{\nu \in M \\ \nu \neq \mathcal{f}}} \mathcal{f} <_M \nu \;\; .$$

By definition of $\mathcal{g}_\nu$ we have further

$$\nu \neq \mathcal{f} \wedge \mathcal{g}_\nu(\underline{A},\underline{D}_\nu,a,b) \wedge \mathcal{g}_{\mathcal{f}}(\underline{A},\underline{D}_{\mathcal{f}},a,b) \rightarrow \neg\lceil \underline{D}_\nu(b) <\!\!-\!\!> \underline{D}_{\mathcal{f}}(b)\rceil \;\; .$$

From this we get easily

$$\nu \neq \mathcal{f} \wedge \mathrm{Min}_{\mathcal{g}_\nu}(\underline{D}_\nu;\underline{A},a,b) \wedge \mathrm{Min}_{\mathcal{g}_{\mathcal{f}}}(\underline{D}_{\mathcal{f}};\underline{A},a,b) \rightarrow$$
$$\rightarrow (\exists x)\lceil a \leq x \leq b \wedge \underline{D}_\nu \tfrac{x}{a} \underline{D}_{\mathcal{f}} \wedge \neg\lceil\underline{D}_\nu(x) <\!\!-\!\!> \underline{D}_{\mathcal{f}}(x)\rceil\rceil \;\; .$$

Together with the formula above we get

$$\bigwedge_{\nu \in M} \mathrm{Min}_{\mathcal{g}_\nu}(\underline{D}_\nu;\underline{A},a,b) \rightarrow \bigvee_{\mathcal{f} \in M} \bigwedge_{\substack{\nu \in M \\ \nu \neq \mathcal{f}}} \mathcal{f} <_M \nu \;\; ,$$

from which follows as in a)

$$\bigwedge_{\nu \in M} \mathrm{Min}_{\mathcal{g}_\nu}(\underline{D}_\nu;\underline{A},a,b) \rightarrow \bigvee_{\nu \in M} \mathrm{Min}_{\mathcal{g}_\nu}(\underline{D}_\nu;\underline{A},a,b') \;\; .$$

Since by induction hypothesis we have

$$(\exists\underline{P}) \,\mathcal{g}_\nu(\underline{A},\underline{P},a,b) \rightarrow (\exists\underline{P})\mathrm{Min}_{\mathcal{g}_\nu}(\underline{P};\underline{A},a,b)$$

and since from (1) follows

$$\overset{m}{\underset{\nu=1}{\bigvee}} \cdot (\exists\underline{P}) \,\mathcal{g}_\nu(\underline{A},\underline{P},a,b) \wedge \mathcal{L}\lfloor\underline{A}(b),Y_\nu,Y_\ell\rfloor \;\; ,$$

we get the wanted conclusion

$$(\exists\underline{P}) \,\mathcal{g}_\ell(\underline{A},\underline{P},a,b') \rightarrow (\exists\underline{P})\mathrm{Min}_{\mathcal{g}_\ell}(\underline{P};\underline{A},a,b'). \quad \#$$

From theorem 1 we get directly the existence of minimal Σ° -predicates in the usual way:

<u>Theorem 2</u>: $(\exists\underline{P}) \,\mathcal{g}(\underline{A},\underline{P},a,b) \rightarrow (\exists\underline{P})\mathrm{Min}_{\mathcal{g}}(\underline{P};\underline{A},a,b)$ is derivable in SC.

Now we are able to solve the problem we started with. Define for $\mathcal{y} = 1, \ldots, m$

$$D_{\mathcal{y}}(d) <\!\!-\!\!> \lceil d < \tilde{b} \wedge (\exists\underline{P}^m)\lceil \mathrm{Min}\,\tilde{\mathfrak{F}}_{1,X,Z}(\underline{P};\underline{B},o,b) \wedge P_{\mathcal{y}}(d)\rceil\rceil \;\vee$$
$$\vee \; (\exists y)(\exists x)\tfrac{y}{b}\lceil H(x) \wedge H(y) \wedge (\exists\underline{P}^{m+1})\lceil \mathrm{Min}\,\tilde{\mathfrak{F}}_{2,Z,Z}(\underline{P};\underline{B},x,y) \wedge P_{\mathcal{y}}(d)\rceil\rceil \;\; .$$

Then $\tilde{\mathcal{f}}(\underline{B},\underline{D})$ follows easily from lemma 1, theorem 2 and the construction of the beginning of the section by a derivation as in the proof of the recursion theorem. The existence of the thus introduced predicates follows by (COMP); thus we conclude $(\exists\underline{P}) \,\tilde{\mathcal{f}}(\underline{B},\underline{P})$, i.e. $\mathcal{f}(\underline{B})$.

c) Syntactization pays

The derivation of section b finishes the completeness proof for SC, the performance of the DP **does** not depend on this derivation. We will now show, however, that the ideas of section b can be used directly to a further considerable improvement of the DP. This improvement allows to replace on the whole the formulae $g_{\nu,\gamma}$ by $\bar{g}_{\nu,\gamma}$, and simplifies thus in turn the completeness proof.

As in section b let $f(\underline{A})$ be a $\Sigma_\wedge^\omega$-formula of length m, let $\bar{g}_{\nu,\gamma}$ be the periodicity formulae corresponding to (cf.4.c). The construction at the beginning of section b suggests to define a set N of pairs of natural numbers by:

$(\nu,\gamma)\epsilon N$ iff there exist numbers $f,\gamma \leq 2^m$ and states $X_0,\dots,X_{f+\gamma}\epsilon 0_m$ such that $\alpha[X_0]$, $X_f \equiv X_{f+\gamma}$, $X_1,\dots,X_f$ are all different, $X_{f+1},\dots,X_{f+\gamma}$ are all different, $\vartheta_{1,X_0,X_1}\epsilon g_\nu$,
$\vartheta_{2,X_f,X_{f+1}}\epsilon g_\gamma$, and for $f = o,\dots,f+\gamma-1$ $\vartheta_{1,X_f,X_{f+1}}\epsilon g_\nu$.

Here $\vartheta\epsilon g$ is short for "ϑ is contained unnegated in g"; the states X_f correspond roughly to the states $\underline{C}(a_\gamma)$ of section b, whereas the numbers $f,f+\gamma$ correspond to f,f respectively. From the existence of such states, in section b we have derived the existence of a carrying predicate for f. This idea leads to

Lemma 1: $\neg f(\underline{A}) <-> \bigvee_{(\nu,\gamma)\not\in N} \bar{g}_{\nu,\gamma}(\underline{A})$

Proof: $->$: Assume $(\nu,\gamma)\epsilon N$, and $\bar{g}_{\nu,\gamma}(\underline{A})$ for fixed $\underline{A}$. We want to show $f(\underline{A})$.
Let f,γ, $X_0,\dots,X_{f+\gamma}$ be as in the definition of N. From $\bar{g}_{\nu,\gamma}(\underline{A})$ follows the existence of a sequence $n_1 < n_2 < \dots$ such that

$g_\nu(\underline{A},o,n_1)$
$g_\gamma(\underline{A},n_\ell,n_{\ell+1})$, $\ell = 1,2,\dots$

This yields by the definition of N
$$\alpha[X_0] \wedge \vartheta_{1,X_0,X_1}(\underline{A},o,n_1) \wedge \bigwedge_{\ell=1}^{f-1} \vartheta_{1,X_\ell,X_{\ell+1}}(\underline{A},n_\ell,n_{\ell+1})$$
and for all $f = o,1,\dots$
$$\vartheta_{2,X_{f+\gamma\cdot f},X_{f+\gamma\cdot f+1}}(\underline{A},n_{f+\gamma\cdot f},n_{f+\gamma\cdot f+1}) \wedge$$
$$\wedge \bigwedge_{\ell=1}^{\gamma-1} \vartheta_{1,X_{f+\gamma\cdot f+\ell},X_{f+\gamma\cdot f+\ell+1}}(\underline{A},n_{f+\gamma\cdot f+\ell},n_{f+\gamma\cdot f+\ell+1}) ,$$

which gives directly $f(\underline{A})$.

Thus we have proved

$$\bigwedge_{(\acute{\iota},\acute{\jmath})\varepsilon N} \lceil \overline{g}_{\acute{\iota}\acute{\jmath}}(\underline{A}) \rightarrow f(\underline{A}) \rceil \quad,$$

from which follows

$$\neg f(\underline{A}) \rightarrow \bigvee_{(\acute{\iota},\acute{\jmath})\notin N} \overline{g}_{\acute{\iota}\acute{\jmath}}$$

by corollary 2.d.2, analogously to end of the proof of lemma 2.d.2.
$\underline{\leftarrow}:$ Let $(\acute{\iota},\acute{\jmath})$ and $\underline{A}$ be fixed, assume $f(\underline{A}) \wedge \overline{g}_{\acute{\iota}\acute{\jmath}}(\underline{A})$. We will show:
$(\acute{\iota},\acute{\jmath})\varepsilon N$.
To this end let the sequence $w_1 < w_2 < \ldots$ be constructed from $g_{\acute{\iota}\acute{\jmath}}(\underline{A})$
as before, let $\underline{C}$ be a carrying predicate for $f(\underline{A})$. Define

$$X_o \equiv_{df} \underline{C}(o) \quad, \quad Y_\ell \equiv_{df} \underline{C}(w_\ell) \quad, \quad \ell = 1,2,\ldots$$

Then

$$\alpha\,[X_o] \wedge \vartheta_{1,X_o,Y_1}(\underline{A},o,w_1) \quad,$$

$$\vartheta_{1,Y_\ell,Y_{\ell+1}}(\underline{A},w_\ell,w_{\ell+1}) \quad, \quad \ell = 1,2,\ldots$$

holds by definition of $\underline{C}$. Since by the definition of the sequence (w_ℓ)

$$h_{\acute{\iota}}(\underline{A},o,w_1) \quad,$$

$$h_{\acute{\jmath}}(\underline{A},w_\ell,w_{\ell+1}) \quad, \quad \ell = 1,2,\ldots$$

is true, we get

$$\vartheta_{1,X_o,Y_1} \varepsilon\, h_{\acute{\iota}}$$

$$\vartheta_{1,Y_\ell,Y_{\ell+1}} \varepsilon\, h_{\acute{\jmath}} \quad, \quad \ell = 1,2,\ldots$$

In view of $(\exists^w t)\,\mathcal{L}\,[\underline{C}(t)]$ there is an infinite sequence $(\ell_{\acute{\jmath}})$ such that

$$\vartheta_{2,Z_{\acute{\jmath}},Z_{\acute{\jmath}+1}}(\underline{A},w_{\acute{\jmath}},w_{\acute{\jmath}+1}) \quad, \quad \acute{\jmath} = 1,2,\ldots$$

is true where $Z_{\acute{\jmath}} \equiv_{df} Y_{\ell_{\acute{\jmath}}}$. This yields as above

$$\vartheta_{2,Z_{\acute{\jmath}},Z_{\acute{\jmath}+1}} \varepsilon\, h_{\acute{\iota}} \quad, \quad \acute{\jmath} = 1,2,\ldots$$

By the finiteness of 0_w there must be a repetition beneath 2^{m+1} in the
sequence $Z_1,Z_2,\ldots;$ let $Z_R \equiv Z_w$ where $R < w \leq 2^m$. Consider the sequence $Y_1,\ldots,Y_{\ell_w}$: If there is a repetition, $Y_{\acute{\jmath}} \equiv Y_\eta$ say,
$1 \leq \acute{\jmath} < \eta \leq \ell_t$, cancel the part $Y_{\acute{\jmath}+1},\ldots,Y_\eta$; go on cancelling, until
there is no repetition between $Y_1,\ldots,Y_{\ell_R}$. The remaining Y_ℓ rename
$X_1,\ldots,X_f$; then $f \leq 2^m$, $X_1 \equiv Y_\ell$, $X_f \equiv Y_{\ell_R} \equiv Z_R$. Do the same with
$Y_{\ell_R+1},\ldots,Y_{\ell_w}$, get $X_{f+1},\ldots,X_{f+g}$, where $g \leq 2^m$, $X_{f+g} \equiv Y_{\ell_w} \equiv X_f$.
Then $X_1,\ldots,X_{f+g}$ satisfies the condition in the definition of N; thus
$(\acute{\iota},\acute{\jmath})\varepsilon N$.

We have derived $(\acute{\iota},\acute{\gamma})\epsilon N$ from $\oint(\underline{A}) \wedge \overline{\mathcal{O}}_{\acute{\iota},\acute{\gamma}}(\underline{A})$, or what amounts to the same, $\overline{\mathcal{O}}_{\acute{\iota},\acute{\gamma}}(\underline{A}) \rightarrow \neg \oint(\underline{A})$ from $(\acute{\iota},\acute{\gamma})\notin N$. I.e. we have shown

$$\bigwedge_{(\acute{\iota},\acute{\gamma})\notin N} \lceil \overline{\mathcal{O}}_{\acute{\iota},\acute{\gamma}}(\underline{A}) \rightarrow \neg\oint(\underline{A}) \rceil \quad .$$

This is equivalent to

$$\bigvee_{(\acute{\iota},\acute{\gamma})\notin N} \overline{\mathcal{O}}_{\acute{\iota},\acute{\gamma}}(\underline{A}) \rightarrow \neg\oint(\underline{A}) \quad .$$

which proves the lemma. $\#$

Lemma 1 allows important changings in part 3 of 4.a, additionally to those noted in 4.c. Namely we can replace steps (5)-(6) by
(6') By inspection of the formulae $\oint_{\acute{\iota}}$, construct the set N defined in 5.c.
Then in step (7) we have to replace ϵM by $\notin N$.

These changings considerably shorten the DP once more. Clearly, the construction of N is yet a serious combinatorial problem; but it is no problem compared with step (6), where a similar problem has to be solved for the much more involved formulae $\oint_{\acute{\iota},\acute{\gamma}}$ and where, moreover, permanent decision on truth of propositional formulae is required. But, what is more essential, for the construction of N the formulae $\overline{\mathcal{O}}_{\acute{\iota},\acute{\gamma}}$ need not be in Σ_1^ω, nor need the formulae $\oint_{\acute{\iota}}$ be in Σ^0. Thus we should best insert step (6') after step (1c) of 4.c: When performing the DP one has just to write down the formulae $\oint_{\acute{\iota}}$ in their original form, and then to construct from them the set N. The remaining steps, which produce lots of predicate quantifiers, have to be performed for the pairs $(\acute{\iota},\acute{\gamma})\epsilon N$ only. Especially, not all formulae $\vartheta_{2,X,Y}$ and $\neg\vartheta_{\acute{\iota},X,Y}$ need to be brought into the form Σ^0.

Besides these improvements of the DP, the definition of N gives rise to a new version of the completeness proof: For the performance of the DP, we replaced in 4.c the formulae $\mathcal{O}_{\acute{\iota},\acute{\gamma}}$ by $\overline{\mathcal{O}}_{\acute{\iota},\acute{\gamma}}$; now we can do the same for the completeness proof, thus the hitch stated in 4.c can be eliminated. Namely it is not difficult to formalize the above proof of lemma 1, thus to derive lemma 1 directly, avoiding the cumbersome lemma 3.b.2.

Lemma 2: $\vdash_{SC} \neg\oint(\underline{A}) \iff \bigvee_{(\acute{\iota},\acute{\gamma})\notin N} \overline{\mathcal{O}}_{\acute{\iota},\acute{\gamma}}(\underline{A})$

Proof: $\rightarrow$: Let $(\acute{\iota},\acute{\gamma})\epsilon N$, let $\oint$, γ, $X_0,\ldots,X_{\oint+\gamma}$ as in the definition of N. For better formulation of the following proof add the formula $\lceil S(t) \wedge S(t') \rightarrow \neg Q(t) \rceil$ as conjunction to the transition condition of $\overline{\mathcal{O}}_{\acute{\iota},\acute{\gamma}}$; this form of $\overline{\mathcal{O}}_{\acute{\iota},\acute{\gamma}}$ is equivalent to the original one.
Assume $\overline{\mathcal{O}}_{\acute{\iota},\acute{\gamma}}(\underline{A})$ to be true, let $\underline{C},\underline{D},E,G$ be carrying predicates for $\overline{\mathcal{O}}_{\acute{\iota},\acute{\gamma}}(\underline{A})$. Define a predicate H by

$$\neg H(o) \ , \quad H(a') <-> \lceil E(a) \wedge \neg E(a')\rceil \vee \lceil \neg E(a) \wedge E(a')\rceil \ .$$

Then H represents the sequence $w_1, w_2, \ldots$ in the proof of lemma 1, i.e.

$$G(a) \wedge \neg G(a') \ -> \ \mathcal{V}_{\dot{\gamma}}(\underline{A}, o, a) \ ,$$

$$a < b \wedge H(a) \wedge (\forall t)_a^b \neg H(t) \wedge H(b) \ -> \ \mathcal{V}_{\dot{\gamma}}(\underline{A}, a, b)$$

is easily derivable for all a. This gives by definition of N

$$G(a) \wedge \neg G(a') \ -> \ \mathcal{A}[X_o] \wedge \mathcal{V}_{1,X_o,X_1}(\underline{A}, o, a) \ ,$$

$$a < b \wedge H(a) \wedge (\forall t)_a^b \neg H(t) \wedge H(b) \ ->$$

$$-> \bigwedge_{\ell=o}^{\dot{t}+y-1} \mathcal{V}_{1,X_\ell,X_{\ell+1}}(\underline{A}, a, b) \wedge \mathcal{V}_{2,X_{\dot{t}},X_{\dot{t}+1}}(\underline{A}, a, b)$$

To derive the existence of a carrying predicate for $f(\underline{A})$ from these formulae, we must find a formal counterpart to the expression "the indices of w_k and w_ℓ are congruent modulo y". To this end we define formulae $t_w(H, a, b)$ whose meaning is "b is the wth element after a in the sequence determined by H":

$$t_o(H, a, b) \equiv_{df} a = b \wedge H(a) \quad ,$$

$$t_1(H, a, b) \equiv_{df} a < b \wedge H(a) \wedge (\forall t)_a^b \neg H(t) \wedge H(b) \quad ,$$

$$t_{w+1}(H, a, b) \equiv_{df} (\exists x) \lceil t_w(H, a, x) \wedge t_1(H, x, b)\rceil \ .$$

With the help of t_w we define formulae $\mathcal{Y}_w(H, a, b)$, for $w = o, \ldots, y-1$, which mean "for some $m \equiv w(y)$, b is the mth element after a in the sequence determined by H"; here y is the number fixed at the beginning of the proof:

$$\mathcal{Y}_w(H, a, b) \equiv_{df} H(a) \wedge (\forall P)\lceil (\forall x) \lceil t_w(H, a, x) \ -> \ P(x)\rceil \wedge$$

$$\wedge \ (\forall x)(\forall y) \lceil t_y(H, x, y) \wedge P(x) \ -> \ P(y)\rceil \ -> \ P(b)\rceil \ .$$

This definition is a generalization and simplification of the formula Od in the proof of theorem 1 in Elgot-Rabin[12], p.172. - Now we can define a carrying predicate for $f(\underline{A})$. Let b and c be numbers so that

$$G(b) \wedge \neg G(b') \wedge t_{\dot{t}}(H, b, c)$$

Then we define for $\dot{\jmath} = 1, \ldots, m$:

$$B_{\dot{\jmath}}(a) <-> \lceil a \leq b \wedge (\exists \underline{P}^m)\lceil Min_{\mathcal{F}_{1,X_o,X_1}} (\underline{P}; \underline{A}, o, b) \wedge P_{\dot{\jmath}}(a)\rceil\rceil \vee$$

$$\vee \ (\exists x)_b^a (\exists y)_a \lceil t_1(H, x, y) \wedge \bigvee_{\ell=1}^{\dot{t}-1} \lceil t_\ell(H, b, y) \wedge$$

$$\wedge \ (\exists \underline{P}^m)\lceil Min_{\mathcal{F}_{1,X_\ell,X_{\ell+1}}} (\underline{P}; \underline{A}, x, y) \wedge P_{\dot{\jmath}}(a)\rceil\rceil\rceil \vee$$

$$\vee \ (\exists x)_c^a (\exists y)_a \lceil t_1(H, x, y) \wedge \bigvee_{\ell=1}^{y-1} \lceil t_\ell(H, c, x) \wedge$$

$$\wedge \ (\exists \underline{P}^{(m)}) \lceil \text{Min } \widetilde{\vartheta}_{1, X_{\not q + \ell}, X_{\not q + \ell + 1}} \quad (\underline{P}; \underline{A}, x, y) \wedge P_{\not q}(a) \rceil \rceil \rceil \ \vee$$

$$\vee \ (\exists x)_c^a (\exists y)_a \lceil \ell_1 (H, x, y) \wedge \not{f}_0 (H, c, x) \wedge$$

$$\wedge \ (\exists \underline{P}^{(m)}) \lceil \text{Min } \widetilde{\vartheta}_{2, X_{\not q}, X_{\not q + 1}} \quad (\underline{P}; \underline{A}, x, y) \wedge P_{\not q}(a) \rceil \rceil \ .$$

Then $\widetilde{\not{f}}(\underline{A}, \underline{B})$ is easily derivable. If we collect the formulae defining $\underline{B}$, $\underline{C}$, $\underline{D}$, E, G, H, b, c and denote their conjunction by $\mathcal{L}$, we have derived:

$$\mathcal{L}(\underline{A}, \underline{B}, \underline{C}, D, E, G, H, b, c) \ \rightarrow \ \widetilde{\not{f}}(\underline{A}, \underline{B})$$

We quantifiy both sides by existential quantifiers; then the left side is derivable from $\overline{\mathcal{G}}_{\acute{\iota}, \acute{\jmath}}(\underline{A})$, and the right side equals $\not{f}(\underline{A})$, which is thus derived from $\overline{\mathcal{G}}_{\acute{\iota}\acute{\jmath}}(\underline{A})$.

<u>←:</u> For the converse direction note that the defining expression of N can be translated into the language of SC; e.g. one has to replace the (bound) existential quantifiers "there are numbers and states" by disjunctions, also the parts $\vartheta_{\acute{\nu}} \ \varepsilon \ \not{f}_{\acute{\jmath}}$ by

$$(\forall \underline{P})(\forall x)(\forall y) \lceil \vartheta_{\acute{\nu}}(\underline{P}, x, y) \ \rightarrow \ \not{f}_{\acute{\jmath}}(\underline{P}, x, y) \rceil$$

and so on. Using this expression for $(\acute{\iota}, \acute{\jmath}) \varepsilon N$, the proof of lemma 1 is easily formalized by methods analogous to those in lemma 5.a.2. #

The reader may wonder how we could avoid lemma 3.b.2 this time, for which - in case of $\overline{\mathcal{G}}_{\acute{\iota}\acute{\jmath}}$ instead of $\mathcal{G}_{\acute{\iota}\acute{\jmath}}$ - we could get no formal proof before. The following lemma 3 shows that in fact we have already derived this lemma. By inspection of the above proof of lemma 2 we would also get a direct proof of lemma 3 not using lemma 2, thus a transformation of section b to the case of formulae $\overline{\mathcal{G}}_{\acute{\iota}\acute{\jmath}}$.

<u>Lemma 3:</u> $\vdash_{SC} \overline{\mathcal{G}}_{\acute{\iota}\acute{\jmath}}(\underline{A}) \wedge \overline{\mathcal{G}}_{\acute{\iota}\acute{\jmath}}(\underline{B}) \ . \rightarrow . \ \not{f}(\underline{A}) \longleftrightarrow \not{f}(\underline{B})$

<u>Proof:</u> Assume $\overline{\mathcal{G}}_{\acute{\iota}\acute{\jmath}}(\underline{A}) \wedge \overline{\mathcal{G}}_{\acute{\iota}\acute{\jmath}}(\underline{B}) \wedge \not{f}(\underline{A})$. The second half of the proof of lemma 1 shows that $(\acute{\iota}, \acute{\jmath}) \varepsilon N$. Then the first half with $\underline{B}$ instead of $\underline{A}$ yields $\not{f}(\underline{B})$. By lemma 2 this proof can be carried through within SC. #

By lemma 2 we can replace the formulae $\mathcal{G}_{\acute{\iota}, \acute{\jmath}}$ by $\overline{\mathcal{G}}_{\acute{\iota}\acute{\jmath}}$ in the whole completeness proof. Theorem 3.b.3 in this form is derived from lemma 2 above and from corollary 3.d.2. Thus in addition to lemma 3.b.2, we spare theorem 3.a.3 (and therefore the whole section 3.a; compare the diagram at the beginning of section a). Moreover, to derive corollary 2.d.2 for $\overline{\mathcal{G}}_{\acute{\iota}\acute{\jmath}}$, we need only a weak Ramsey theorem which we get from theorem 1.c.3 by adding $\wedge \ (\forall t)_x^y \neg Q(t)$ to the clause $Q(x) \wedge Q(y)$. This weakening allows corresponding alterations in the proof of the lemmata proving theorem 1.c.2. But these alterations do not change the course

of the proof; thus the completeness proof is not earnestly simplified
what regards the Ramsey theorem. Since the derivation of lemma 2 uses
minimal Σ^{0}-predicates just as does section b to derive lemma 3.b.2,
what we really spare is section a.3.

Nevertheless, one may ask why we introduced the formulae $\mathcal{G}_{\nu,\gamma}$ at all,
instead of working directly with the formulae $\overline{\mathcal{G}}_{\nu,\gamma}$. Now, in the here
presented version we derived the Ramsey theorem to demonstrate the
power of SC. Later on the consideration on negated $\Sigma^{\omega}_{\Lambda}$-formulae led
wholly naturally to the periodicity formulae $\mathcal{G}_{\nu,\gamma}$ and to lemma 2.d.2,
from which we got theorem 3.b.3, and thus the decidability of SC, by
the Ramsey theorem. Only when we looked at the DP for feasibility we
saw that the formulae $\mathcal{G}_{\nu,\gamma}$, though clear in notation and meaning, are
too clumsy to be used in the DP. Thus we replaced them within the DP
by the formula $\overline{\mathcal{G}}_{\nu,\gamma}$; on the other hand, the formalization of the proof
of the auxillary lemma 3.b.2 encountered serious difficulties, and
seemed to be impossible in case of $\overline{\mathcal{G}}_{\nu,\gamma}$. But it was just the analysis of
those difficulties which made more clear the relation between a $\Sigma^{\omega}_{\Lambda}$-for-
mula f and its associated periodicity formulae $\mathcal{G}_{\nu,\gamma}$ resp. $\overline{\mathcal{G}}_{\nu,\gamma}$. By this
we found a direct proof for theorem 3.b.3 in case of $\overline{\mathcal{G}}_{\nu,\gamma}$ which led to a
considerable improvement of the DP on one hand, and to a simplification
of the completeness proof on the other hand.

The first attempt to translate the DP of Büchi into the language of
SC has led to Σ^{0}- and Σ^{0}_{R}-formulae as compensation for finite automata,
and has made the DP more complicated. But when the syntactic version
was at hand, it turned out that a formal system, rigid though it be, is
a very good tool in dealing with a DP. It is just the extreme punctiliousness of a fixed formal system which shows the points where the DP
is too involved to be performed; the same punctiliousness, however,
allows to express things in different ways, and thus to choose always
the best suited formulation. So by the attempt to get a complete axiom
system from a DP, one is forced to make inferences more and more elementary (sc. monadic second order elementary), and thus more and more
complicated. This way down the steps of the DP becomes more elementary,
too, and thus more easy; moreover it gives insight into the possibilities of the system, and thus into possibilities for easier transformations. — The way from lemma 2.d.2 to lemma 2 of this section seemed to
be a good illustration for these claims; therefore we did not present
at once the final form of the DP and of the completeness proof, but we
built it up step by step.

Chapter II.   Benefits of the decision procedure

As emphazised in the introduction, a DP is slightly more than a crib
for a lazy mathematician to prove his theorems. Indeed, if the system
is not wholly trivial, in most cases it is much too difficult to carry
through a DP. The usefulness of a DP is to be found rather in the information about the system in question obtained from the DP. So we have
got in chapter I a complete axiom system for SC consulting the DP. In
this chapter we will use the DP to answer questions concerning the
strength of SC: What relations on natural numbers are expressible in
SC? And what about relations on sets of natural numbers? In other words:
To which part of the whole number theory does SC correspond? This question is answered best by comparing SC with other number theoretic systems, and by asking for the there definable sets and relations.

Thus we will treat in §1 and §2 of this chapter the problem of definability by formulae containing resp. free predicate variables and free
individual variables. 1.a is concerned with definable sets of words.
The true connection is given between SC and the theory of automata;
further the concepts of regular and multiperiodic sets are introduced.
1.b deals with definable sets of threads, and with implicitely and explicitely definable threads; the problem of eliminating quantifiers in
formulae with free predicate variables is attacked, but not solved.
1.b is used in 1.c to look into a hierarchy Σ_M, Π_M of formulae defined
analogously to Σ_M^ω, Π_M^ω. Just as 1.b does for threads, 2.a shows that
exactly the ultimately periodic relations on natural numbers are defi-
nable in SC; but to this end one has to modify for higher dimensions
the concept "ultimately periodic". From this definability result follows (2.b) that SC is equivalent to the elementary theory of congruence
and order (which yields the effective quantifier elimination for formulae without free predicate variables); moreover, it allows to characterize also the definable functions and wellorderings. 1.b and 2.a are
used in 2.c to ascertain the standard models of SC.

More light is thrown on the strength of SC by comparing it in §3
with related systems known from the literature: with weak second order
arithmetic in 3.a and with a similar system in 3.b. In 3.c we extend
the results from the natural numbers to the integers.

§1. Definitions involving free predicate variables

a) Finite automata, recursion, and definable sets of words

Several times we have alluded to the strong connection between SC
and the theory of finite automata. Now we will set up this connection
precisely, in showing that Σ_{R}^{o}- and Σ^{o}-formulae are in some sense
equivalent to resp. deterministic and nondeterministic finite automata.
To settle the questions of notation we start with a short account of the
notion of a finite automaton, following Rabin-Scott[30]. For further information we refer to this paper and to the literature (cf.e.g.Church[10]
and McNaughton[23]).

The exact definition of a finite automaton is based on the following
intuitive idea: Let a finite set 0, the "alphabet", be given, the elements of which are called "letters". Then an "automaton" Υ is a system,
capable of finitely many "states", which is able to read letters; any
feeded letter changes the state of Υ into another one in a predescribed
manner. The behaviour of Υ is thus determined by the "transition function" f which attaches a state to any pair (state, letter). Obviously,
also every finite sequence of letters ("tape") changes the automaton
from the given state into another one: the automaton reads letter by
letter changing its states each time. We may describe this process by
the flow diagram:

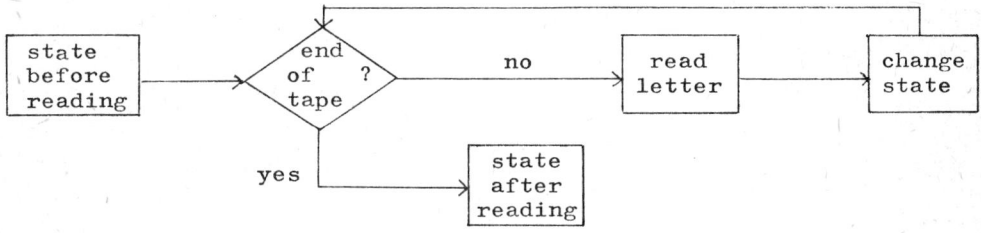

Among the states an "initial state" and "final states" are distinguished.
The automaton "accepts" a tape if it reads the tape starting from the
initial state and stops at a final state.

We chose as alphabet the set O_{n} (n be fixed in the following), and
may thus speak of words instead of tapes. (For unexplained notation see
I.2.a.)

Definition 1: A (deterministic finite n-) automaton is a quadrupel
$\Upsilon = (S, f, s_{o}, S_{1})$, where S is a finite non-void set (the states of Υ),
$f : S \times O_{n} \longrightarrow S$ is a function (the transition function of Υ), $s_{o} \varepsilon S$, and
$S_{1} \subseteq S$ (the initial state resp. the final states of Υ). (Υ means "se-

quential machine", another name for finite automata).

We extend f to a function $f : S \times T_w \longrightarrow S$ by the recursion $f(s,\wedge) = s$, $f(s,uX) = f(f(s,u),X)$ $(s\varepsilon S,\ u\varepsilon T_w,\ X\varepsilon O_w)$. Thus if γ starts on s reading u , it stops on $f(s,u)$.

<u>Definition 2:</u> γ <u>accepts an w-word</u> u, if $f(s_o,u)\varepsilon S_1$. $\underline{T(\gamma)}$ is the set of words accepted by γ.

In his papers $[8],[9]$ and $[10]$, Church handles automata problems by "wider restricted recursive arithmetic" - a numbertheoretic system similar to SC which has no quantifiers, but allows the introduction of predicate constants by a certain recursion rule. Written in our no- tation and shortened by the conventions of I.1.a, a simple instance of this recursion schema would read as

$$(RR) \qquad \begin{array}{l} \underline{B}^w(o) <-> Z^w \\[2mm] \underline{B}^w(a') <-> \underline{\mathcal{L}}^w[\underline{A}^w(a),\underline{B}(a)] \end{array}$$

The "new" predicate constants $B_1,\ldots,B_w$ are introduced by simultaneous recursion involving the values of the "old" predicate constants $A_1,\ldots$ $\ldots,A_w$. Since Church has slightly another concept of automaton, he con- siders a more general schema, but this one is just good to compare re- cursions and Rabin-Scott automata. To this end, think an w-automaton $\gamma = (S,f,s_o,S_1)$ be given (without any loss of generality) by a finite set of elements each capable of exactly two values T and F. Thus, if γ consists of w elements, any state of γ is described by an w-tuple of T's and F's; consequently γ has 2^w states. Now it is self-evident how to describe γ by recursions: Identify S with the set O_w (not to be confused with the alphabet O_w); thus s_o is an element Z of O_w , S_1 a subset of O_w , and $f = (f_1,\ldots,f_w)$ a function from $O_w \times O_w$ to O_w. Let the predicate variables $B_1,\ldots,B_w$ correspond to the elements of γ, and let propositional formulae $\underline{\mathcal{L}}^w$ and $\mathcal{L}$ correspond to resp. the transition function and the final states. More exactly, by the method of the expanded normal form of the propositional calculus, construct $w+1$ propositional formulae $\mathcal{L}[\underline{H}^w]$, $\mathcal{L}_i[\underline{G}^w,\underline{H}^w]$ $(i = 1,\ldots,w ;\ G_i,\ H_i$ propositional variables) such that $\mathcal{L}[Y]$ is true exactly for $Y\varepsilon S_1$ and $\mathcal{L}_i[X,Y]$ is true if and only if $f_i(Y,X) = T$. Let u be a word of length l, let $\underline{A}$ be any w-tuple of predicates, the initial part of which equals u. Then γ accepts u if and only if the schema (RR) - speci- fied by these Z, $\underline{\mathcal{L}}$,$\underline{A}$ - yields a value of $\underline{B}(l)$ such that $\mathcal{L}[\underline{B}(l)]$ is true. In this way, to any automaton corresponds a triple $(Z, \underline{\mathcal{L}}, \mathcal{L})$ which "accepts" the same words; the converse is trivial. In $[2]$, Büchi calls these triples "finite automata".

Now the correspondence between SC and automata theory is obvious. In
I.2.a, recursion led to the definition of Σ_R^o-formulae, called "recursive Σ^o-formulae". And the one-to-one-correspondence between the
triples $(Z, \underline{\mathcal{b}}, \mathcal{L})$ above and Σ_R^o-formulae

$$(\exists \underline{P}^{\mathcal{W}}).\left\lceil \underline{P}(a) <-> Z\right\rceil \wedge (\forall t)_a^b\left\lceil \underline{P}(t') <-> \underline{\mathcal{b}}\left[\underline{A}^{\mathcal{W}}(t),\underline{P}(t)\right]\right\rceil \wedge \mathcal{L}\left\lfloor \underline{P}(b)\right\rfloor$$

is trivial. In other words we have shown: If $\gamma = (S,f,s_o,S_1)$ is a
finite deterministic automaton over the alphabet $O_{\mathcal{W}}$, then one can construct effectively a Σ_R^o-formula which accepts the same words as γ.
Clearly, this correspondence is anticipated by the terminology for
Σ^o-formulae. A Σ^o-formula "accepts" a word, as an automaton accepts
a tape; even the letter "T" in $T_{\mathcal{W}}$ and $T(\mathcal{f})$ reminds us on "tapes".

For better formulation we introduce the following terminology (cf.
I.2.a):

<u>Definitions</u>: 1) A Σ^o-formula $\mathcal{f}$ <u>defines a set of words</u> $U \subseteq T_{\mathcal{W}}$ iff
$U = T(\mathcal{f})$. Then U is a <u>definable set of words</u>. $\underline{DT_{\mathcal{W}}}$ is the set of all definable subsets of $T_{\mathcal{W}}$.
2) <u>A finite automaton</u> γ <u>defines a set of words</u> $U \subseteq T_{\mathcal{W}}$ iff $U = T(\gamma)$.
Then U is an <u>automaton definable set of words</u>. $\underline{ADT_{\mathcal{W}}}$ is the set of all
automaton definable subsets of $T_{\mathcal{W}}$.
3) A Σ^o-formula $\mathcal{f}(\underline{A}^{\mathcal{W}})$ and an $\mathcal{W}$-automaton γ are <u>equivalent</u> if they determine equal sets of $\mathcal{W}$-words: $T(\mathcal{f}) = T(\gamma)$. Analogously two automata, or
two Σ^o-formulae are <u>equivalent</u>.

By theorem I.2.c.3, $DT_{\mathcal{W}}$ is closed under intersection, union and complement, thus is a Boolean subalgebra of $\pi(T_{\mathcal{W}})$ under these operations.
$DT_{\mathcal{W}}$ is obviously countable, and therefore a non-trivial subalgebra of
$\pi(T_{\mathcal{W}})$, i.e. one not coinciding with $\pi(T_{\mathcal{W}})$ (for an example see below).
Remark that $\mathcal{f}$ is equivalent to $\mathcal{g}$ by definition 3 iff $\mathcal{f}(\underline{A},a,b) <->$
$<-> \mathcal{g}(\underline{A},a,b)$ is derivable in SC. Remark further that each of these
equivalence relations is decidable (Corollary I.2.c.1; Rabin-Scott[30],
corollary 10.1).

Now the above result reads as follows: To every finite automaton one
can construct effectively an equivalent Σ_R^o-formula. The converse is
trivial: Let $\mathcal{f}$ be a Σ_R^o-formula as above, let $\mathcal{R} =_{df} L(\mathcal{f})$ (cf.I.4.b).
Then $\gamma =_{df} (O_{\mathcal{R}},f,Z,S_1)$ is equivalent to $\mathcal{f}$, where $f(Y,X) =_{df} \underline{\mathcal{b}}[X,Y]$
and $S_1 =_{df} \{Y; \mathcal{L}[Y]\}$.
Since by theorem I.2.c.2 any Σ^o-formula is equivalent to a Σ_R^o-for-
mula, we have shown:

<u>Theorem 1</u> (Büchi[3], lemma 2): $DT_{\mathcal{W}} = ADT_{\mathcal{W}}$, i.e. the same sets of words
are definable by Σ^o-formulae and by finite automata. Moreover, from
any Σ^o-formula one can construct effectively an equivalent finite

automaton, and conversely.

As a consequence of theorem 1 we note an example of a set of words not definable in SC. Namely, by Rabin-Scott[30] the set $U =_{df} \{F^{m} TF^{m} ; \; m = 0,1,2,..\}$ is not automaton definable. That $U \notin DT_1$ follows also directly by methods as used in the proof of theorem b.3: A process cannot transmit information over an arbitrary intervall of time.

Büchi[3] does not use Σ_R^o-formulae. Thus at points where we change from Σ^o to Σ_R^o, he has to evade to finite automata by the above theorem 1. There is on the other hand a generalization of the above concept of finite automata to so-called "nondeterministic" finite automata; nondeterministic automata correspond to Σ^o-formulae as exactly as do deterministic automata to Σ_R^o-formulae. A Σ_R^o-process − as we called the carrying predicate of a Σ_R^o-formula − is determined uniquely by its beginning state and its transition recursion (i.e. by the formula) from the received information (i.e. from the predicate making the formula true). Σ^o-processes on the contrary are not fully determined, they can vary within given conditions. Similarly, the action of a nondeterministic automaton is not fully determined by its state and the feeded letter; also the initial state is not fully determined.

Definition 4: A <u>nondeterministic</u> (finite w-) automaton is a quadrupel $\gamma = (S,f,S_o,S_1)$, where S is a finite set (the <u>states</u> of γ), $f : S \times O_w \longrightarrow \pi(S)$ is a function into the set of all subsets of S (the <u>transition function</u> of γ), and $S_o, S_1 \subseteq S$ (the <u>initial</u> resp. <u>final</u> <u>states</u> of γ).

If a nondeterministic automaton reads a word u then in general at each step there will be many possible directions for the automaton to move on, in accordance with the transition function; i.e. there will be a lot of ways through the states of the automaton which all can result from reading u. It is natural to say that the automaton accepts u if among these possible ways there is one which starts in S_o and ends in S_1.

Definition 5: A <u>nondeterministic w-automaton</u> $\gamma = (S,f,S_o,S_1)$ <u>accepts</u> <u>an w-word</u> u of length ℓ, if there is a sequence $s_o,\ldots,s_\ell$ of states of γ such that

(1) $s_o \in S_o$

(2) $s_{\nu+1} \in f(s_\nu,u(\nu))$, $\nu = 0,\ldots,\ell-1$

(3) $s_\ell \in S_1$

The analogy between nondeterministic automata and Σ^o-formula is obvious. The fact that Σ^o-formulae are in some sense not more general than Σ^o_R-formulae, corresponds to the reduction of nondeterministic automata to deterministic ones (Rabin-Scott[30], theorem 11). Indeed, the proof of our theorem I.2.c.2 uses the same ideas as the proof of Rabin-Scott, which again is likely the "subset-construction of Myhill[25]", a paper to which Büchi[3],p.4 refers for a proof of his lemma 2, but which was not available for us. Conversely we can show that any nondeterministic automaton is equivalent to a deterministic one, using theorem I.2.c.2 and the correspondence between nondeterministic automata and Σ^o-formulae.

To be sincere, we have carried through the above construction which showed the correspondence between deterministic automata and Σ^o_R-formulae, only in the case where γ is an automaton consisting of m elements each of them capable of two values T and F. The generalization is trivial: Let $\gamma = (S, f, S_o, S_1)$ be any nondeterministic n-automaton, we want to construct an equivalent Σ^o-formula. (The argument for deterministic automata and Σ^o_R-formulae is similar, but a little less easy to write down.) Let $s_1, \ldots, s_\ell$ be the states of γ. Choose R so that $\ell \leq 2^R$, fix a numbering $X_1, \ldots, X_{2^n}$ and $Y_1, \ldots, Y_{2^R}$ of O_n and O_R, resp. Define sets $K =_{df} \{\nu; s_\nu \varepsilon S_o\}$, $L =_{df} \{(\nu, j, \ell); \ s_j \varepsilon f(d_j, X_\nu)\}$, $M =_{df} \{\nu; s_\nu \varepsilon S_1\}$. Then

$$(\exists \underline{P}^R). \bigvee_{\nu \varepsilon K} \lceil \underline{P}(a) <\!\!-\!\!> Y_\nu \rceil \wedge$$

$$\wedge \bigvee_{(\nu, j, \ell) \varepsilon L} \lceil \lceil \underline{A}(t) <\!\!-\!\!> X_\nu \rceil \wedge \lceil \underline{P}(t) <\!\!-\!\!> Y_j \rceil \wedge \lceil \underline{P}(t') <\!\!-\!\!> Y_\ell \rceil \rceil \wedge$$

$$\wedge \bigvee_{\nu \varepsilon M} \lceil \underline{P}(b) <\!\!-\!\!> Y_\nu \rceil$$

is equivalent to γ. The construction is in fact the same as before, we have only to identify the states $s_1, \ldots, s_\ell$ and the tuples $Y_1, \ldots, Y_\ell$.

The comparison with finite automata yields yet another information about SC: In [13], Elgot and Wright consider elementary arithmetic making use only of the ordering relation and of one-place predicate variables. They call a formula "singular" if it contains exactly one free individual variable (and arbitrary many free predicate variables), and show that singular formulae cannot define all automaton definable sets of words. This indicates the increasing of strength by adding predicate quantifiers: in SC, all automaton definable sets are definable by the "singular" formulae $\oint(\underline{A}, o, a)$ of Σ^o.

For sake of completeness we will give now - following mostly the presentation of Rabin-Scott[30] - two other, pure mathematical characterizations of the definable word sets. The first is due to Kleene and Myhill:

Definition 6: Let U,V be subsets of T_w :

a) $\underline{UV} =_{df} \{uv; u\epsilon U, v\epsilon V\}$ is the <u>complex product</u> of U and V.

b) $m > 0$: $\underline{U^m} =_{df} \underbrace{UU...U}_{m}$; $\underline{U^0} =_{df} \{\wedge\}$.

c) $\underline{cl(U)} =_{df} \bigcup_{m=0}^{\infty} U^m$ is the <u>closure</u> of U.

Thus words in UV arise from affixing a word of V to a word of U, words in cl(U) arise from concatenating finitely many words of U.

Definition 7: The concept of a <u>regular set of w-words</u> is defined by the following recursion:

The finite subsets of T_w are regular. If $U,V \subseteq T_w$ are regular, then $U \cup V$, UV, and cl(U) are regular, too.

It is well known that the regular sets are just the sets definable by finite automata. Therefore, in view of theorem 1 we need not show that the class of regular sets coincides with the class DT_w. It is easy, however, and very instructive to prove directly that the regular sets are definable by Σ^0-formulae. We suggest this task as an exercise to the interested reader. (Hint: to show that DT is closed under formation of closure use the idea manifested in the formulae $\overline{q}_{ij}$ of I.4.c.) - From the fact that all one-element word sets are definable by Σ^0-formulae follows that DT_w is an atomic Boolean algebra. Moreover, DT_w is closed under the operations involved into definition 7. (Surely it would be interesting to study the general properties of Boolean algebras on semigroups.)

The second characterization is due to Myhill:

Definition 8: Let R be an equivalence relation on T_w:

a) R is a <u>congruence</u> if it is compatible with concatenation, i.e. if for any words u,v,w R(u,v) implies R(uw,vw) and R(wu,wv).

b) The <u>rank</u> of R is the cardinal number of its equivalence classes.

The definition of congruence is the usual one if we regard T as a semigroup under concatenation (cf.I.2.a).

Definition 9: A subset U of T_w is <u>multi-periodic</u> if it is closed with respect to some congruence of finite rank.

The concept of a multi-periodic set of words should not be confused with our notion of a "thread multi-periodic from Σ^0-formulae g and ζ" of I.2.d; at best, the use of this term in I.2.d could explain the choice of the term "multi-periodic" for the above concept. - Theorem 1 of Rabin-Scott[30] tells that the multi-periodic sets coincide with the sets definable by finite automata. Again it is easy and very in-

structive to translate the proof of the quoted theorem into a proof
which shows directly that exactly the multi-periodic sets are definable
by Σ^o-formulae. For the first half of the proof one has to define a
Σ^o_R-formula similar to that in the proof of theorem I.2.c.2. For the
second half, one has to use the idea which led in I.2.d to a congruence
of finite rank induced by a Σ^ω_1-formula. As we remarked there, Büchi
uses just the above theorem of Rabin-Scott to get defining automata
(and thus Σ^o-formulae) for the classes of the quoted congruence. Again
we will leave the translation to the reader.

We conclude this section by a collection of the quoted results, thus
extending theorem 1:

<u>Theorem 2:</u> Let U be a subset of T_ν. The following assertions are "effec-
tively" equivalent

(1) U is definable in SC by a Σ^o_R-formula.

(2) U is definable in SC by a Σ^o-formula.

(3) U is definable by a determinisitic finite automaton.

(4) U is definable by a nondeterminisitic finite automaton.

(5) U is definable in W2A.

(6) U is definable in L^3_1.

(7) U is regular.

(8) U is multi-periodic.

(5) and (6) anticipate results of 2.a and 2.b.

b) Definable sets of threads and definable threads

The preceding section motivated the terminology used in connection with Σ°-formulae: Σ°-formulae are the SC-compensation for finite automata. We got Σ°-formulae in I.2.a as Σ_1^{ω}-formulae which stop working somewhere; conversely we may regard Σ_1^{ω}-formulae as finite automata which work at infinity. This is the point of view of Elgot-Rabin [12], section 2: A finite automaton "accepts an infinite tape" iff there is an infinite sequence of states of the automaton which satisfies the initial condition, and infinitely often the final condition, and which is "compatible" with the transition condition and the given tape. In exactly the same sense a Σ_1^{ω}-formula "accepts a thread" according to the definition in I.2.a; our "carrying thread" is the "compatible sequence of states". Since Σ_1^{ω} is a normal form for SC-formulae without free individual variables, in I.2.a we introduced this terminology for arbitrary such formulae.

In this section we will look for what kind of sets of threads come into the picture, if we consider such formulae and the threads accepted by it. For terminology we refer to I.2.a; the letter "S" in "S " and "S(f)" means "snakes", i.e. "infinitely long words", and has no connection with automata theory.

Definition 1: An <u>SC-formula</u> f without free individual variables <u>defines a set of threads</u> $\phi \subseteq S_{w}$ iff $\phi = S(f)$. Then ϕ is a <u>definable set of threads</u>. DS_{w} is the set of all definable subsets of S_{w}.

Let w be fixed. The theorems 1-3 of I.3.b tell that the class DS_{w} of definable sets of threads is a (countable) Boolean algebra. Moreover, the proof of lemma I.2.d.2 gives a complete characterization of DS_{w} in terms of multi-periodic sets: Let $\phi \in DS$ be defined by the formula $f(\underline{A}^{w})$. In view of theorem I.4.a.1 we can assume f to be a Σ_1^{ω}-formula. Let formulae $\gamma_i \in \Sigma^{\circ}$, $i = 1, \ldots, \ell$, and $g_{i,\gamma}$, $i, \gamma = 1, \ldots, \ell$, be defined from f as in I.2.d. From the proof of lemma I.2.d.2 follows easily

$$(1) \quad f(\underline{A}) \longleftrightarrow \bigvee_{(i,\gamma) \notin M} g_{i,\gamma}(\underline{A})$$

Now the formulae $g_{i,\gamma}$ are just defined to accept exactly the threads of the form $uv_1 v_2 v_3 \ldots$ where u satisfies γ_i and the v_γ for arbitrary γ satisfy γ_{γ}. Thus, using the notation of I.2.a and of the preceding section, we may write

$$S(g_{i,\gamma}) = T(\gamma_i)T(\gamma_{\gamma})T(\gamma_{\gamma})T(\gamma_{\gamma})\ldots.$$

Since further, by (1),

$$S(f) = \bigcup_{(i,\gamma) \notin M} S(g_{i,\gamma})$$

and since the sets $T(\gamma_i)$ are multi-periodic, we get

Theorem 1 (Büchi[3], lemma 10): Let ϕ be a subset of S_N: ϕ is in DS_N if and only if there exist a number ℓ, a subset M of $\ell \times \ell$, and multi-periodic sets U_i, $i = 1,\ldots,\ell$, so that

$$\phi = \bigcup_{(i,j)\in M} U_i U_j U_j U_j \ldots$$

Moreover, if a formula f defining ϕ is given, then ℓ,M, and Σ°-formulae γ_i defining U_i can effectively be constructed. Conversely, ℓ,M and multi-periodic sets U_i determine effectively a formula f defining ϕ.

To get the last statement of the theorem, use theorem a.2 to construct formulae γ_i defining U_i, and then compose formulae $g_{i,j}$ from γ_i as predescribed by M. - Clearly one could use in the above proof the formulae $\bar{g}_{i,j}$ and the set N of I.5.c instead of $g_{i,j}$ and M; note that the representation of ϕ need not be the same in both cases.

Theorem 1 yields another proof of corollary I.2.b.4, of which we give now a more general version (for terminology see definition 1 in I.2.b):

Corollary 1: Any definable set ϕ of threads which is not empty, contains an ultimately periodic thread. From a formula f defining ϕ one can construct effectively examples of ultimately periodic threads satisfying f.

Of special interest are the definable singleton sets:

Definition 2: An SC-formula defines a thread φ iff it defines the set $\{\varphi\}$ consisting of φ alone. A Σ°-formula defines a word u iff it defines $\{u\}$.

Thus, $f(\underline{A})$ defines a thread, iff the "existence and unicity formula"

$$(\exists\underline{P}):f(\underline{P}) \wedge (\forall\underline{R}).f(\underline{R}) \to (\forall x)\lceil\underline{P}(x) <\to \underline{R}(x)\rceil$$

is derivable.

Corollary 2: Just the ultimately periodic threads are definable in SC. Initial part and periodic germ of a defined thread can be computed effectively from the defining formula.

Proof: That the definable threads are ultimately periodic follows immediately from corollary 1. That the converse is also true is easily seen from the proof of theorem 1: Construct formulae γ_i and γ_j which define resp. the non-periodic part and the periodic germ of the wanted thread; then $\bar{g}_{i,j}$ defines this thread. The effectivity statement follows from corollary 1, too. #

It follows that DS_N is, like DT_N, an atomic Boolean algebra, which nevertheless does not contain every one-point set, since not every thread is ultimately periodic. - Note that we cannot simply replace $\overline{\mathcal{G}}_{i,\gamma}$ by $\mathcal{G}_{i,\gamma}$ in the above proof. Namely, if $\mathcal{G}_{i,\gamma}$ is satisfiable, the incorporated formula $\mathcal{H}_\gamma$ cannot define a word at all, indeed $T(\mathcal{H}_\gamma)$ must be infinite. For let $\mathcal{G}_{i,\gamma}$ be satisfied by $\underline{A}$, let B be a carrying thread for $\mathcal{G}_{i,\gamma}(\underline{A})$; then

$$(\exists^\omega t)B(t) \wedge (\forall y)(\forall x)_0^y \Big[B(x) \wedge B(y) \to \mathcal{H}_\gamma(\underline{A},x,y) \Big]$$

is true. Thus $\underline{A}(a)\ldots\underline{A}(b-1)\epsilon T(\mathcal{H}_\gamma)$ for any pair a < b from B. Thus if we replace in the above proof $\mathcal{H}_\gamma$ by a formula $\mathcal{H}_\gamma^*$ which accepts, not only the periodic germ u , but also any repetition u...u of it but no other words, then we may choose $\mathcal{G}_{i,\gamma}$ instead of $\overline{\mathcal{G}}_{i,\gamma}$ at will. This example shows that $T(\mathcal{H}_\gamma)$ may well be infinite, but $\mathcal{G}_{i,\gamma}$ defines a thread. Remember also the example of I.4.c, where $\mathcal{H}_\gamma$ accepted just the 1-words containing an odd number of F's, and therefore $\mathcal{G}_{i,\gamma}(A)$ was not satisfiable whereas $\overline{\mathcal{G}}_{i,\gamma}(A)$ was equivalent to $(\exists^\omega t)\neg A(t)$.

Instead of using formulae $\mathcal{G}_{i,\gamma}$ we can define any ultimately periodic thread explicitly:

Definition 3: An SC-formula $\oint$ defines explicitly a thread iff $\oint$ is of the form

(1) $(\forall t)\Big[\underline{A}(t) <\to \mathcal{O}\!\mathcal{I}(t)\Big]$,

where neither $\underline{A}$ nor other free predicate variables are contained in $\mathcal{O}\!\mathcal{I}$.

Note that for formulae of form (1) existence and unicity are always derivable; indeed the existence formula is a special case of (COMP). Therefore any explicitly definable thread is definable. The converse is by no means obvious, but also true. For let $\underline{A}^*$ be definable; by corollary 2, $\underline{A}$ is ultimately periodic. So let $\underline{A}$ be of the form uvvv..., let $u \equiv X_0 X_1 \ldots X_{m-1}$, $v \equiv Y_0 Y_1 \ldots Y_{\ell-1}$, define $\mathcal{O}\!\mathcal{I}_\gamma(a)$, $\gamma = 1,\ldots,m$, as

(2) $\displaystyle\bigvee_{i=0}^{m-1} \Big[a = i \wedge (X_i)_\gamma \Big] \vee \Big[m \leq a \wedge \bigvee_{i=0}^{\ell-1} \Big[a \equiv m + i(\ell) \wedge (Y_i)_\gamma \Big] \Big]$

where $(X_i)_\gamma$ is the γth component of the tuple X_i. Then clearly $\underline{A}$ and $\mathcal{O}\!\mathcal{I}$ satisfy (1). Since corollary 2 is effective, we conclude:

Corollary 3: Any thread definable in SC is explicitly definable. The formulae $\mathcal{O}\!\mathcal{I}$ of the equivalence (1) can be computed effectively from the defining formula.

Let CO be the elementary theory of congruence and order over the natural numbers, i.e. the theory formalized in first order predicate calculus using o, ', <, and $\equiv (n)$, $n = 1,2,\ldots$ as primitive constants, fixed by appropriate axioms. Then CO is a subtheory of the elementary

theory P of addition over natural numbers, since zero, successor,
order, and congruence are definable in P; or to be more precisely: CO
is interpretable in P. The elementary theory of addition was first
developed for the integers in Presburger [26]; there he showed the
completeness of an axiom system by proving that this theory is decidable. A reformulation for the natural numbers – which we call P to
remind on "Presburger" – is to be found in Hilbert-Bernays [20], vol.
I,p.357, together with a decidability and completeness proof. (For information on elementary theories, especially on interpretability, see
Tarski-Mostowski-Robinson [41]; see also section 2.c and [15])

Get COP from CO by adding one-place predicate variables, but no predicate quantifiers, to CO. Then COP is interpretable in SC, and the
above proof shows that corollary 3 can be sharpened to

Corollary 3a: Any thread definable in SC is explicitely definable in
COP (or even, if one regards the definiens alone, in CO).

This result seems to make redundand our recursion theorem I.1.b.1.
The recursion theorem states that any formula of the form

(R) $\underline{E}(a) \longleftrightarrow \underline{\mathcal{L}}(a,\underline{E}(t); \ t < a)$

defines a thread, i.e. that existence and unicity are derivable for
such formulae. If now any "definition" (R) can be converted into an
explicit one,

(3) $\underline{E}(a) \longleftrightarrow \underline{\mathcal{Q}}(a)$ ($\underline{E}$ not in $\underline{\mathcal{Q}}$) ,

then the existence of a predicate defined by (R) seems to be derivable
in SC directly with the help of (COMP) without using the recursion
theorem. But this is not true, since we need the completeness of our
axiom system to derive in SC that the predicates defined by (R) and
(3) respectively are the same, i.e. to derive

$$(\forall x)\lceil \underline{E}(x) \longleftrightarrow \underline{\mathcal{Q}}(x)\rceil \wedge (\forall x)\lceil \underline{D}(x) \longleftrightarrow \underline{\mathcal{L}}(x,\underline{D})\rceil \rightarrow$$
$$\rightarrow (\forall x)\lceil \underline{E}(x) \longleftrightarrow \underline{D}(x)\rceil \ .$$

Corollary 3a can be stated in another form. First of all, we could
replace in (2) the part $w \leq a$ by $\overset{w-1}{\underset{\nu=0}{\wedge}} a \neq \nu$, and thus dispense with
order. But we are interested more in the fact that the formula (2)
does not contain quantifiers at all, be it predicate or individual
quantifiers. Thus we have

Corollary 3b: To any SC-formula which defines a thread one can construct effectively an equivalent SC-formula of the form $(\forall t) \mathcal{L}(t)$ where
$\mathcal{L}$ is quantifier-free.

By the decidability proof of Presburger [26], the theory P allows
effective quantifier elimination (i.e. to any formula one can construct
effectively an equivalent one which is quantifier-free), if one adds $<$
and $\equiv (\mathcal{w})$, $\mathcal{w} = 1,2,\ldots$ as primitive constants; from the proof follows
that also CO allows effective quantifier elimination. In 2.a we will
show that for any SC-formula without free predicate variables one can
construct effectively an equivalent CO-formula; thus SC allows effective quantifier elimination for such formulae. Here we have to look
for an analogous result for formulae containing free predicate (but no
free individual) variables. For such formulae corollary 3b above is the
best we can hope: Since in SC-formulae a predicate variable must always
occur with an individual term attached, such formulae are bound to contain at least one individual quantifier. But corollary 3 is formulated
for very special formulae and it seems questionable whether it can be
generalized, i.e. whether the clause "which defines a thread" can be
replaced by "without free individual variables". To do so one would
have to determine the elements of the Boolean algebra $DS_{\mathcal{w}}$ by its atomic
elements.

The beginning of the section suggests another direction of investigation: Theorem 1 (or more directly lemma I.2.d.2), and in a much easier
way corollary I.2.b.3, show that Σ_1^{ω}-formulae, and thus arbitrary SC-
formulae without free individual variables, can be expressed smoothly
with the help of Σ°-formulae. Thus one has to look for quantifier elimination of such formulae. Again the example of I.4.c makes clear the
difficulties. There we constructed a very simple formula $\gamma_4(A,a,b)$ of
the meaning "A has an odd number of F's between a and b" what seems
not to be expressible in COP. It may well be that CQP is weaker than
SC, in the sense that it cannot provide an equivalent formula for any
SC-formula (without free individual variables). In this case one has to
extend COP by new primitives, e.g. by one of the meaning of $\gamma_4(A,a,b)$,
or by another of the meaning "A is ultimately true, and has an odd number of F's in its non-constant initial part". It will be shown in 4.b
that it is always possible to extend a theory in such a way that the
extension allows quantifier elimination, and that it is decidable if
the original theory was. But this procedure is by no means effective,
and it would be very elaborate to try it in case of SC.

We conclude this section by infering from corollary 1 a result concerning non-definability:

Corollary 4: The property of being ultimately periodic is not definable
in SC, nor is the property of being periodic. I.e.: there is no formula
$\{(\underline{A}^{n})$ such that $S(\{)$ is the set of the ultimately periodic resp. the

periodic ω-threads.

Proof: Assume f to define the set of ultimately periodic threads. Then the set of threads not ultimately periodic would be defined by $\neg f$. Thus $S(\neg f)$ would contain no ultimately periodic thread, in contrary to corollary 1. – Assume further f to define the set of periodic threads. Then

$$(\exists P).f(P) \wedge (\exists x)(\forall y)_x \lceil P(y) <\rightarrow A(y) \rceil$$

would accept just the ultimately periodic threads, which is impossible by the first half of the proof. #

c) The hierarchy Σ_w, Π_w

In I.1.d we defined the hierarchy Σ_w^ω, Π_w^ω, $w = 1,2,\ldots$ of classes of formulae, and showed that Σ_w^ω is a normal form for SC-formulae without free predicate variables, i.e. that any such formula is contained in Σ_w^ω for some w. The proof gave a slightly stronger result, namely that we could replace the quantifier $(\exists^\omega t)$ in the final condition of Σ_w^ω by $(\exists t)$. We will now use section b to consider the hierarchy Σ_w, Π_w which results from thus altering the final conditions; especially we will show that we could not use Σ_w instead of Σ_w^ω in I.1.d, since Σ_1 is not closed under negation, and thus is not a normal form. Büchi uses the hierarchy Σ_w, Π_w in [3]; we will partly extend, and partly correct his results.

Definition 1: Let Σ_w be the class of formulae which result from Σ_w^ω-formulae by replacing $(\exists^\omega t)$ by $(\exists t)$ in the final condition. Analogously define Π_w from Π_w^ω. Let $\widetilde{\Sigma}_w$ and $\widetilde{\Pi}_w$ be the classes of formulae equivalent to Σ_w- resp. Π_w-formulae.

First of all we show that both Σ_2 and Π_2 are normal forms.

Lemma 1: Σ_w and Π_w are closed under conjunction and disjunction.

Proof: For Σ_1, the proof carries over nearly unchanged from the corresponding theorems concerning Σ_w^ω, I.3.b.1+2. For Σ_w, $w > 1$, one has first to restitute the prefix in the right way. The proof for Π_w is similar. #

Theorem 1: To any formula without free individual variables one can construct effectively an equivalent one in Π_2. The same holds if one replaces Π_2 by Σ_2.

Proof: Let $\int(\underline{A})$ be a formula without free individual variables. By theorem I.4.a.1, we may $\neg\int$ assume to be equivalent to a Σ_1^ω-formula, and therefore

(1) $\int(\underline{A}) :\longleftrightarrow: (\forall\underline{P}).\, \alpha(o) \vee (\exists t)\, \mathcal{G}(t) \vee (\forall^\omega t)\, \mathcal{L}(t)$.

Using once more a switching predicate one sees easily that

$(\forall^\omega t)H(t) :\longleftrightarrow: (\exists Q).Q(o) \wedge (\forall t)\lceil \neg Q(t) \rightarrow \neg Q(t') \wedge H(t)\rceil \wedge$
$$\wedge (\exists t)\neg Q(t) \quad.$$

If we insert this into (1), we get the formula

(2) $(\forall\underline{P}): \alpha(o) \vee (\exists t)\, \mathcal{G}(t) \vee (\exists Q).Q(o) \wedge (\forall t)\lceil \neg Q(t) \rightarrow \neg Q(t') \wedge \mathcal{L}(t)\rceil \wedge$
$$\wedge (\exists t)\neg Q(t)$$

equivalent to $\int$. Regarding each of the three disjuncts as an "impure"

(cf.I.3.a) Σ_1-formula, we transform (2) into a Σ_2-formula $\mathcal{G}$ by applying lemma 1 twice. Impurity may be removed from $\mathcal{G}$ by introducing
auxiliary predicate variables, as at the end of the proof of theorem
I.1.d.1. Thus we have shown the second half of the theorem; the first
half follows by interchanging $\int$ and $\neg\int$ in the above proof. $\#$

  Next we show that theorem 2 cannot be improved:

Theorem 2: Let $\phi =_{df} S((\exists^\omega t)A(t))$, $\Psi =_{df} S((\forall^\omega t)\neg A(t))$ be the sets of
infinite and finite threads, resp. Then ϕ is definable in π_1, but not
in Σ_1; on the other hand, Ψ is definable in Σ_1, but not in π_1.

Proof: From the beginning of the preceding proof follows that Ψ is definable in Σ_1, and thus ϕ, in π_1. Assume now ϕ to be defined by the
Σ_1 -formula

$$\mathcal{G}(A) \equiv (\exists \underline{P}^\omega).\, \mathcal{U}[\underline{P}(o)] \wedge (\forall t)\mathcal{L}[A(t),\underline{P}(t),\underline{P}(t')] \wedge (\exists t)\mathcal{L}[\underline{P}(t)] .$$

For $\mathcal{G}(A)$ to be true it suffices the final condition $\mathcal{L}(t)$ to be satisfied once and never more. We will use this fact to construct a finite
thread which satisfies $\mathcal{G}$, in contrary to the assumption. Since we have
to do with formulae, we will carry out the construction speaking of predicates instead of threads.

Let $\ell =_{df} 2^\omega$, define a predicate A by

$$A(a) <\!\!-\!\!> a \equiv o \; (\ell+1) .$$

A satisfies $\mathcal{G}$. Let $\underline{B}^\omega$ be a carrying predicate of $\mathcal{G}(A)$, let a number
a be chosen so that

$$\mathcal{U}[\underline{B}(o)] \wedge (\forall t)\mathcal{L}[A(t),\underline{B}(t),\underline{B}(t')] \wedge \mathcal{L}[\underline{B}(a)] .$$

Let b be a number so that

$$a \leq b \wedge b \equiv o \; (\ell+1)$$

By the usual argument there must be numbers i, j , $1 \leq i < j \leq \ell+1$,
so that

$$\underline{B}(b+i) \equiv \underline{B}(b+j)$$

From this we infer

$$(\exists \underline{P}).\,\lceil \underline{P}(i) <\!\!-\!\!> X\rceil \wedge (\forall t)_i^j \,\mathcal{L}[F,\underline{P}(t),\underline{P}(t')] \wedge \lceil \underline{P}(j) <\!\!-\!\!> X\rceil ,$$

where $X \equiv_{df} \underline{B}(b+i)$. From this follows

$$(\exists \underline{P}).\,\lceil \underline{P}(o) <\!\!-\!\!> X\rceil \wedge (\forall t)\mathcal{L}[F,\underline{P}(t),\underline{P}(t')] .$$

Define a word w and predicates C and $\underline{D}^\omega$ by

$$w \equiv_{df} \underline{B}(b+i+1)\underline{B}(b+i+2)...\underline{B}(b+j-1),$$

$$C \equiv_{df} A(o)A(1)...A(b)FFF...$$

$$\underline{D} \equiv_{df} \underline{B}(o)\underline{B}(1)\ldots\underline{B}(b+\iota-1)XwXwXw\ldots$$

Then $\underline{D}(a) \equiv \underline{B}(a)$, thus

$$\alpha[\underline{D}(o)] \wedge (\forall t)\mathcal{L}[C(t),\underline{D}(t),\underline{D}(t')] \wedge \mathcal{L}[\underline{D}(a)] .$$

This implies $\mathcal{G}(C)$, which is a contradiction since $(\forall^\omega t)\neg C(t)$. #

Corollary 1: Σ_1 and π_1 are not closed with respect to negation.

For the following theorem 3, which summarizes the foregoing discussion, compare remark 2 at p.7 of Büchi[3], the second half of which is wrong. Replacing within this remark Σ_2 and π_2 by Σ_1 and π_1, resp., and cancelling the word "much" we get

Theorem 3:

a) $\quad \tilde{\Sigma}_w = \tilde{\pi}_w = \tilde{\Sigma}_1^\omega$ for $w \geq 2$.

b) $\quad \tilde{\Sigma}_1 \overset{c}{\neq} \tilde{\pi}_2$, $\tilde{\pi}_1 \overset{c}{\neq} \tilde{\Sigma}_2$

c) $\quad \tilde{\Sigma}_1 \neq \tilde{\pi}_1$

Thus for $w \geq 2$ both, Σ_w and π_w, are a normal form for SC-formulae: the hierarchy Σ_w, π_w collapses above $w = 1$. But the beginning steps are genuine:

Σ_1 and π_1 are different from each other and from the whole set of SC-formulae.

§2. Definitions involving free individual variables

Except for the case of Σ^o-formulae, where the free individual variables served as dummy variables, we have not yet considered formulae containing free individual variables. We will show in this section that again just the ultimately periodic sets of natural numbers are definable by those formulae which contain one free individual variable. To extend this result to relations we have to distinguish several concepts of "ultimately periodic". From the so obtained classification of the definable relations, in section b we get the effective quantifier elimination for SC-formulae without free predicate variables; moreover we characterize the definable functions and well-orderings. In section c, we consider the standard models of SC.

a) Definable sets of numbers

Analogous to the case of threads and words we call an n-tuple of numbers an n-number. An n-set is a set of n-numbers. - Let f be a formula with n free individual variables:

Definition 1: $N(f)$ is the set of all n-numbers which satisfy f. $N \subseteq \mathbb{N}^n$ is a definable subset of $\mathbb{N}^n$ iff $N = N(f)$ for some f.

To examine the definable sets we need a simple normal form for formulae containing free individual variables. And indeed, theorem I.4.a.1 gives us a Σ_1^ω-like normal form for those formulae, too:

Theorem 1 (Büchi[3], p.7, remark 3): To any formula $f(\underline{A}, \underline{a}^n)$ one can construct effectively an equivalent one of the form $\Sigma_{1,n}^\omega$

$$(\exists \underline{P}). \, \alpha[\underline{P}(o)] \wedge (\forall t) \, \mathcal{L}[\underline{A}(t), \underline{P}(t), \underline{P}(t')] \wedge (\exists^\omega t) \mathcal{L}[\underline{P}(t)] \wedge$$
$$\wedge \bigwedge_{\nu=1}^{n} \vartheta_\nu [\underline{P}(a_\nu)] \quad .$$

Proof: Let $f(\underline{A}, \underline{a}^n)$ be given. Following the proof of the corresponding theorem 3 in Büchi[2] we eliminate the free individual variables in f with the help of switching predicates, and apply then theorem I.4.a.1: Define:

$$f_1(\underline{A}, \underline{B}^n) \equiv_{df} \bigwedge_{\nu=1}^{n} B_\nu(o) \wedge (\forall t)\left[\underline{B}(t') \to \underline{B}(t)\right] \wedge$$
$$\wedge (\forall \underline{x}^n) \left\lceil \bigwedge_{\nu=1}^{n} \left\lceil B_\nu(x_\nu) \wedge \neg B_\nu(x_\nu') \right\rceil \to f(\underline{A}, \underline{x}) \right\rceil \quad .$$

Then holds

$$(1) \quad f(\underline{A}, \underline{a}) :<->: (\exists \underline{P}^n). \, f_1(\underline{A}, \underline{P}) \wedge \bigwedge_{\nu=1}^{n} \left\lceil P_\nu(a_\nu) \wedge \neg P_\nu(a_\nu') \right\rceil \quad .$$

Eliminating the unwanted arguments a_ν' we get

(2) $f(\underline{A},\underline{a}) :<->: (\exists\underline{P}^w\underline{R}^w)\cdot f_1(\underline{A},\underline{P}) \wedge (\forall t)\lceil\underline{R}(t) <-> \underline{P}(t')\rceil \wedge$

$$\wedge \bigwedge_{\nu=1}^{w} \lceil P_\nu(a_\nu) \wedge \neg R_\nu(a_\nu)\rceil \quad .$$

Since f_1 does not contain free individual variables we can construct, using theorem I.4.a.1, an equivalent Σ_1^ω-formula f_2 of the form:

$$(\exists\underline{S})\cdot\alpha[\underline{S}(o)] \wedge (\forall t)\mathcal{B}[\underline{A}(t),\underline{B}(t),\underline{S}(t),\underline{S}(t')] \wedge (\exists^\omega t)\mathcal{L}[\underline{S}(t)] \quad .$$

Replacing in (2) $f_1(\underline{A},\underline{B})$ by $f_2(\underline{A},\underline{B})$ we get the wanted result

$f(\underline{A},\underline{a}) :<->: (\exists\underline{PRS})\cdot\alpha[\underline{S}(o)] \wedge$

$\wedge (\forall t)\lceil\mathcal{B}[\underline{A}(t),\underline{P}(t),\underline{S}(t),\underline{S}(t')] \wedge \lceil\underline{R}(t) <-> \underline{P}(t')\rceil\rceil \wedge$

$\wedge (\exists^\omega t)\mathcal{L}[\underline{S}(t)] \wedge \bigwedge_{\nu=1}^{w} \lceil P_\nu(a_\nu) \wedge \neg R_\nu(a_\nu)\rceil \quad . \quad \#$

Now we carry over the definition of "ultimately periodic" from w-threads to w-sets. To this end we extend the tuple convention of I.1.a with respect to numbers: Given any number m, $\overline{m}^w$ (or simply $\overline{m}$) denotes the "constant" w-tuple $(m,\ldots,m)$. As before, however, $\underline{m}$ denotes an arbitrary w-tuple, the components of which are $m_1,\ldots,m_w$. Analogous to the case of recursion equivalences we shorten w-tuples of equations, unequalities, congruences, and addition and multiplication terms by writing e.g. $\underline{m} + \overline{\underline{z}}\,\underline{a} \equiv \underline{\ell}(\gamma)$ instead of $\bigwedge_{\nu=1}^{w} m_\nu + z a_\nu \equiv \ell_\nu(\gamma)$.

Definition 2: An w-set N is <u>ultimately periodic</u> (<u>u.p.</u> for short) iff there are numbers f and $z > 0$ so that for every w-number $\underline{m}$ where $\overline{f}^w \leq \underline{m}$, $\underline{m} \in N$ if and only if $\underline{m}+\overline{\underline{z}}\in N$. f and z are resp. the <u>phase</u> and the <u>period</u> of N.

Considering first the case $w = 1$ we see easily that any definable set of numbers is u.p.: Let $N = N(f)$, define

(3) $\mathcal{G}(A) \equiv_{df} (\forall t)\lceil A(t) <-> f(t)\rceil \quad .$

Clearly there exists exactly one thread φ satisfying $\mathcal{G}$, which is u.p. in view of corollary 1.b.2. Since φ is the indicator of N (i.e. is true for just the elements of N), we get N to be u.p. – Conversely, any u.p. set N of phase f and period z is definable by the SC-formula

(4) $\bigvee_{m\in F} a = m \vee \lceil f \leq a \wedge \lceil\bigvee_{m\in P} a \equiv m (z)\rceil\rceil$

where $F =_{df} \{m\in N; m < f\}$ and $P =_{df} \{m\in N; f \leq m < f+z\}$.

Instead of this, of course we could go again the round-about-way through corollary 1.b.2 to construct a formula $\mathcal{G}(A)$ which defines A as the indicator of N; then the formula

(5) $(\forall P)\lceil\mathcal{G}(P) \to P(a)\rceil$

defines N. Thus we conclude

<u>Remark 1:</u> By formulae containing one free individual variable, exactly the ultimately periodic sets are definable.

The above proof shows, either directly, or by its first half combined with corollary 1.b.3a, that to any SC-formula with one free individual variable and no free predicate variables one can construct effectively an equivalent one in CO which is quantifier-free. It will be the main result of this section that the same is true for formulae with more free individual variables.

When we turn now to the case $w = 2$ the picture changes totally. Since SC has only one-place predicates, the strong connection is lost between formulae containing free predicate variables and formulae containing free individual variables. A 2-thread is stratified using two independent sequences, whereas a 2-set is really two-dimensional. Surely we may extend the above formula (5) to $(\forall PQ)\lceil \mathcal{U}(P,Q) \rightarrow P(a) \wedge Q(b)\rceil$; but the 2-set thus defined gives a very distorted image of the 2-thread defined by $\mathcal{U}$. And extending similarly the above formula (3) yields a formula

$$(\forall x)(\forall y)\lceil A(x) \wedge B(y) <-> \mathcal{f}(x,y)\rceil$$

which needs not even define any 2-thread.

Therefore we have to show directly - without using section 1.b - that any definable 2-set is u.p., and conversely. - Attacking the second task first, we generalize the formula (4) to

(6) $\quad \bigvee_{\underline{m}\,\epsilon\, F} \underline{a} = \underline{m} \vee \lceil \overline{\mathcal{f}} \leq \underline{a} \wedge \bigvee_{\underline{m}\,\epsilon\, P} \underline{a} \equiv \underline{m} (\mathcal{z})\rceil$,

where $F =_{df} \{\underline{m}\epsilon N;\ \underline{m} < \overline{\mathcal{f}}\}$, $P =_{df} \{\underline{m}\epsilon N;\ \overline{\mathcal{f}} \leq \underline{m} < \overline{\mathcal{f}} + \overline{\mathcal{z}}\}$.
But formula (6) is not the intended one, since it is false whenever

$\quad a_1 < \mathcal{f} \leq a_2 \vee a_2 < \mathcal{f} \leq a_1 .$

Thus we replace in (6) $\overline{\mathcal{f}} \leq \underline{a}$ by $\mathcal{f} \leq a_1 \vee \mathcal{f} \leq a_2$, getting the formula (7). Clearly any 2-set N defined by a formula of type (7) is u.p., but the converse is not true. This is best seen by comparing the following two pictures:

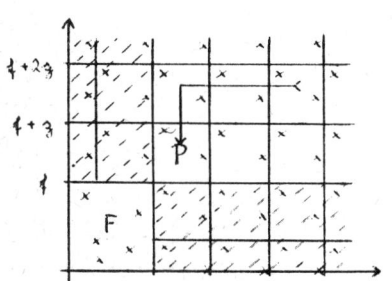
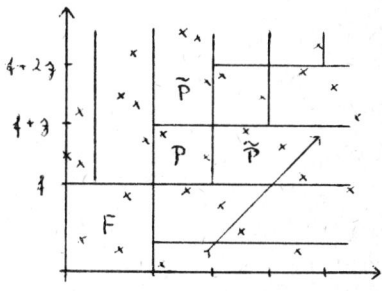

Pict.1: N is u.p. in layers   Pict.2: N is u.p.

In both pictures elements of N are represented by crosslets in the plane. Consider now picture 1: Any square with the exception of F is congruent to P. Thus we may test any point $\underline{a}^2$ of the plane whether it belongs to N, by shifting it first horizontally to the column of P (asking: $\bigvee_{\underline{w} \varepsilon P} a_2 \equiv \underline{w}_2 (\gamma)$?) and then pushing it down (or lifting it) to P (asking: $\bigvee_{\underline{w} \varepsilon P} a_1 \equiv \underline{w}_1 (\gamma)$?). $\underline{a}^2$ belongs to N if and only if both answers are positive. (The striped region is added to N by passing from (6) to (7)). - In contrary, if N is any ultimately periodic set we are allowed to shift points only in the diagonal (Pict.2). This is expressed by replacing in (7) $\bigvee_{\underline{w} \varepsilon P} \underline{a} = \underline{w} (\gamma)$ (which is short for $(\exists \underline{x}^2) \underline{a} \pm \overline{\gamma} \underline{x} = \underline{w}$) by $(\exists x) \underline{a} \pm \overline{\gamma} \overline{x} = \underline{w}$. Further we have to replace P by $\widetilde{P} = \{\underline{w} \varepsilon N; \phi \leq \underline{w}_1 < \phi + \gamma \lor \phi \leq \underline{w}_2 < \phi + \gamma \}$, since otherwise N is empty outside the squares in the diagonal of pict.2. But this is impossible, since $\widetilde{P}$ is infinite. - Now we give an exact definition of the property illustrated in pict.1:

Definition 3: An n-set N is <u>ultimately periodic in layers</u> iff there are numbers ϕ and $\gamma > 0$ (resp. the <u>phase</u> and the <u>period</u> of N) such that: for every n-number $\underline{w}$ where $\phi \leq \underline{w}_\iota$ for at least one ι, $\underline{w} \varepsilon N$ iff $(\underline{w}_1, \ldots, \underline{w}_{\gamma-1}, \underline{w}_{\gamma+\gamma}, \underline{w}_{\gamma+1}, \ldots, \underline{w}_n) \varepsilon N$ for all γ, $1 \leq \gamma \leq n$.

We have shown: Any 2-set (and similarly any n-set) u.p. in layers is definable in SC. That the converse is not true shows e.g. the formula a = b. We shall show on the other hand that any definable n-set is u.p. That the converse is not true is suggested by the above consideration: In general, an u.p. set is not reducible to finitely many (initial) conditions. Thus we have to look for a property of n-sets which lies between the both considered.

To this end let the formula $\oint(\underline{a}^2)$ (without free predicate variables) be given, and recall the equivalence (1) in the proof of theorem 1. If we replace $\oint_1(\underline{B})$ by $\oint_2(\underline{B})$ in (1) instead of (2), we get the following formula as being equivalent to $\oint$:

$$(\exists \underline{PS}). \mathcal{O}[\underline{S}(o)] \land (\forall t) \mathcal{G}[\underline{P}(t), \underline{S}(t), \underline{S}(t')] \land (\exists^\omega t) \mathcal{L}[\underline{S}(t)] \land$$
$$\land \bigwedge_{\iota=1}^{2} \lceil P_\iota(a_\iota) \land \neg P_\iota(a_\iota') \rceil .$$

Thus the 2-number $\underline{w}$ is contained in $N(\oint)$ if and only if there is a thread φ in $S(\oint_2)$ such that $\varphi_\iota(\underline{w}_\iota)$ is true and $\varphi_\iota(\underline{w}_\iota + 1)$ is false for $\iota = 1,2$ (where φ_ι is the ιth component of φ). Let φ be a thread of $S(\oint_2)$: By theorem 1.b.1 there are Σ^o-formula γ_1 and γ_2 so that $\varphi \equiv uv_1v_2v_3\ldots$, where $u \varepsilon T(\gamma_1)$, $v_\iota \varepsilon T(\gamma_2)$, $\iota = 1,2,\ldots$. Let $\underline{w}$ be a 2-number, let $\varphi(\underline{w}_1)$ be a state of v_γ, say, and $\varphi(\underline{w}_2)$ a state of v_ℓ, where e.g. $\gamma < \ell$. If in φ we replace v_1 by v_γ and v_2 by v_ℓ, we get a

33

new thread ψ, which again satisfies $\mathfrak{f}_2$ (cf.I.2.d). Thus we have found:
To see whether $\underline{m} \varepsilon N$ it suffices to test a certain initial portion of
any thread φ satisfying $\mathfrak{f}_2$, by shifting down $\varphi(\underline{m}_1)$ and $\varphi(\underline{m}_2)$ in their
order to this region. For φ can be replaced by an u.p. thread which
satisfies $\mathfrak{f}$, too, and has this initial portion as non-periodic part and
periodic germ. This is of course a very rough account, but it will perhaps make better understandable the following definition. – On the
other hand, if we can give in advance an upper bound of the length of
the said initial portions to be tested, we can get an SC-definition of
ultimately periodic 2-sets of this sort.

<u>Definition 4</u>: A 2-set N is <u>fanlike ultimately periodic</u> if there are
numbers $\mathfrak{f}$ and $\mathfrak{z}>0$ (resp. the <u>phase</u> and the <u>period</u> of N) so that for every
2-number $\underline{m}$, $\underline{m} \varepsilon N$ if and only if one of the following five conditions
is fulfilled (where $L =_{df} \{\underline{\ell} \varepsilon N;\ \lceil \ell_1 < \mathfrak{f} + \mathfrak{z} \wedge 0 \le \ell_2 - \ell_1 < \mathfrak{f} + \mathfrak{z} \rceil \vee$

$\vee \lceil \ell_2 < \mathfrak{f} + \mathfrak{z} \wedge 0 \le \ell_1 - \ell_2 < \mathfrak{f} + \mathfrak{z} \rceil \}$):

(i) $\underline{m} \varepsilon L$

(ii) $\mathfrak{f} + \mathfrak{z} \le \underline{m}_1 \le \underline{m}_2$ and $(\mathcal{R}, \underline{m}_2 - \underline{m}_1 + \mathcal{R}) \varepsilon N$ where $\mathcal{R}$ is so that

     $\mathfrak{f} \le \mathcal{R} < \mathfrak{f} + \mathfrak{z}$ and $\underline{m}_1 \equiv \mathcal{R} \ (\mathfrak{z})$

(iii) $\mathfrak{f} + \mathfrak{z} \le \underline{m}_2 - \underline{m}_1$ and $(\underline{m}_1, \underline{m}_1 + \mathcal{R}) \varepsilon N$ where $\mathcal{R}$ is so that

     $\mathfrak{f} \le \mathcal{R} < \mathfrak{f} + \mathfrak{z}$ and $\underline{m}_2 \equiv \underline{m}_1 + \mathcal{R} \ (\mathfrak{z})$

(iv)   and (v) like (ii) resp. (iii) where $\underline{m}_1$ and $\underline{m}_2$ are interchanged.

That a so defined set is periodic in some sense, is suggested by regarding the congruences modulo $\mathfrak{z}$ in the definition. The very structure
of a fanlike u.p. set is realized best by considering the following
picture:

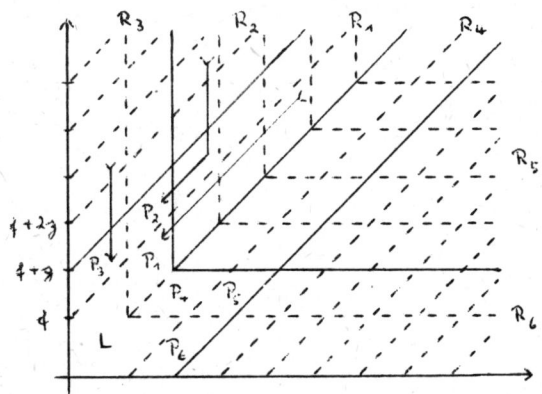

Pict.3: N is fanlike u.p.

Again N is represented as a subset of the plane (for better survey
we deleted the crosslets indicating the elements of N). To any of the
five cases of the definition correspond certain of the regions L,
$R_1,\ldots,R_6$, which are bordered by a continuous line. $R_1 \cup R_2$ corresponds
to (ii), $R_2 \cup R_3$ to (iii); analogous correspond $R_4 - R_6$ to (iv) and (v).
Corresponding to (i) there is the "initial portion" L of N, consisting
of the "periodic germ" and the "non-periodic part". The periodic germ
is composed out of six (disjoint) subsets $P_1 - P_6$. Any region is built
up by repetitions of certains of the P_ν: R_1 from $P_1 \cup P_2$, R_2 from P_2,
R_3 from $P_2 \cup P_3$; analogously $R_4 - R_6$ from $P_4 - P_6$. For R_1 and R_3 this
follows directly from the definition, for R_3 it is seen by combining
the conditions (ii) and (iii) (see the proof of lemmata 2 and 3). For
these three cases we have indicated by an arrow the "motion of a point
into his testing region" (cf. the proof of lemmata 1 and 2 below). -
An easy example of a fanlike u.p. 2-set is the set N = $\left\{ \underline{m}^2; \; m_1 < m_2 \right\}$
of phase 1 and period 1.

Now we have to show that fanlike u.p. sets are just what we need:
Lemma 1: Any 2-set definable in SC is fanlike ultimately periodic.
Proof: Let N = N(f) be a 2-set where, by theorem 1, f may be assumed to
have the $\Sigma_{1,2}^\omega$-form

$$(\exists \underline{P}^\gamma).\, \alpha[\underline{P}(o)] \wedge (\forall t)\, \mathcal{L}[\underline{P}(t),\underline{P}(t')] \wedge (\exists^\omega t)\, \mathcal{L}[\underline{P}(t)] \wedge \overset{2}{\underset{\nu=1}{\wedge}}\, \gamma_\nu[\underline{P}(a_\nu)] \; .$$

As in the proof of theorem I.2.b.2 we decompose the formula f into
several formulae which impose conditions on certain parts of the carry-
ing predicates of f: Let $f =_{df} 2^\gamma$, let $Y_1,\ldots,Y_f$ be the elements of O_γ,
define for $\nu, \gamma = 1,\ldots, f$

$$\vartheta_{\nu\gamma}(a,b) \equiv_{df} (\exists \underline{P}^\gamma).\, \lceil \underline{P}(a) <\!\!-\!\!> Y_\nu \rceil \wedge (\forall t)_a^b \, \mathcal{L}[\underline{P}(t),\underline{P}(t')] \wedge \lceil \underline{P}(b) <\!\!-\!\!> Y_\gamma \rceil$$

$$\vartheta_\nu^\omega(a) \equiv_{df} (\exists \underline{P}^\gamma).\, \lceil \underline{P}(a) <\!\!-\!\!> Y_\nu \rceil \wedge (\forall t)_a \, \mathcal{L}[\underline{P}(t),\underline{P}(t')] \wedge (\exists^\omega t)\, \mathcal{L}[\underline{P}(t)]$$

Let $a \leq b$: Then $f(a,b)$ holds if and only if there are states Y_ν, Y_γ, Y_ℓ
so that

$$\alpha[Y_\nu] \wedge \gamma_1[Y_\gamma] \wedge \gamma_2[Y_\ell] \wedge \vartheta_{\nu\gamma}(o,a) \wedge \vartheta_{\gamma\ell}(a,b) \wedge \vartheta_\ell^\omega(b)$$

Define K $=_{df} \left\{ (\nu,\gamma,\ell,\kappa); \; \alpha[Y_\nu] \wedge \gamma_1[Y_\gamma] \wedge \gamma_2[Y_\ell] \wedge \vartheta_\kappa^\omega(o) \right\}$
Then

$$f(a,b) <\!\!-\!\!> \lceil a \leq b \wedge \underset{(\nu,\gamma,\ell,\ell)\in K}{\vee} \lceil \vartheta_{\nu\gamma}(o,a) \wedge \vartheta_{\gamma\ell}(a,b) \rceil \rceil \;\; \vee$$

$$\vee \lceil b \leq a \wedge \underset{(\nu,\gamma,\ell,\gamma)\in K}{\vee} \lceil \vartheta_{\nu\ell}(o,b) \wedge \vartheta_{\ell\gamma}(b,a) \rceil \rceil \;\; .$$

The $\vartheta_{\nu\gamma}$'s are Σ_2^0-formulae; let $f_{\nu\gamma}$ and $z_{\nu\gamma}$ be the phase resp. the
period of $\vartheta_{\nu\gamma}$ for $\nu, \gamma = 1,\ldots, f$. Let $f =_{df} \underset{\nu\gamma}{\max}(f_{\nu\gamma})$, let z be the

least common multiple of the $\gamma_{\nu,\gamma}$. Any $\vartheta_{\nu,\gamma}$ may be regarded as having
phase ϕ and period γ. We want to show: N is fanlike ultimately periodic
of phase ϕ and period γ. Thus define a set L as in definition 4; let
$\underline{m} \, \varepsilon N$, let $m_1 \leq m_2$:

1.case: $\underline{m} \, \varepsilon L$. Ready.

2.case: $\phi + \gamma \leq m_1 \wedge o \leq m_2 - m_1 < \phi + \gamma$. Let κ be the number so
that $\phi \leq \kappa < \phi + \gamma \wedge m_1 \equiv \kappa \ (\gamma)$. Then $\vartheta_{\nu,\gamma}(o, m_1) <-> \vartheta_{\nu,\gamma}(o, \kappa)$ for
any ν, γ by corollary I.2.b.2. Further $\vartheta_{\gamma,\ell} (m_1, m_2) <-> \vartheta_{\gamma,\ell} (\kappa, m_2 - m_1 + \kappa)$
for any γ, ℓ by lemma I.2.b.2. Therefore

$$\phi (m_1, m_2) <-> \phi (\kappa, m_2 - m_1 + \kappa)$$

3.case: $\phi + \gamma \leq m_2 - m_1$. Let κ be the number so that
$\phi \leq \kappa < \phi + \gamma \wedge m_1 + \kappa \equiv m_2 \ (\gamma)$. Again by corollary I.2.b.2,
$\vartheta_{\gamma,\ell}(m_1, m_2) <-> \vartheta_{\gamma,\ell}(m_1, m_1 + \kappa)$ for any γ, ℓ and therefore
$\phi(m_1, m_2) <-> \phi(m_1, m_1 + \kappa)$. — If $m_2 \leq m_1$, the argument is analogous;
the lemma is proved. #

<u>Lemma 2:</u> Any fanlike ultimately periodic 2-set is definable in SC.

<u>Proof:</u> Let N be such a set, let ϕ and γ be its phase and period, resp.
Define the formula ϕ (a,b) as

$$\bigvee_{\underline{\ell} \varepsilon L} (a,b) = \underline{\ell} \quad \vee$$

$$\vee \ulcorner \phi + \gamma \leq a \wedge b < a + \phi + \gamma \wedge$$

$$\wedge \bigvee_{\kappa_1 = \phi}^{\phi + \gamma - 1} \bigvee_{\kappa_2 = 0}^{\phi + \gamma - 1} \ulcorner a \equiv \kappa_1 (\gamma) \wedge a + \kappa_2 = b \wedge \bigvee_{\underline{\ell} \varepsilon L} (\kappa_1, \kappa_1 + \kappa_2) = \underline{\ell} \urcorner \urcorner \quad \vee$$

$$\vee \ulcorner a < \phi + \gamma \wedge a + \phi + \gamma \leq b \wedge$$

$$\wedge \bigvee_{\kappa = \phi}^{\phi + \gamma - 1} \ulcorner b \equiv a + \kappa (\gamma) \wedge \bigvee_{\underline{\ell} \varepsilon L} (a, a + \kappa) = \underline{\ell} \urcorner \urcorner \quad \vee$$

$$\vee \ulcorner \phi + \gamma \leq a \wedge a + \phi + \gamma \leq b \wedge$$

$$\wedge \bigvee_{\kappa_1, \kappa_2 = \phi}^{\phi + \gamma - 1} \ulcorner (a,b) \equiv (\kappa_1, \kappa_1 + \kappa_2)(\gamma) \wedge \bigvee_{\underline{\ell} \varepsilon L} (\kappa_1, \kappa_1 + \kappa_2) = \underline{\ell} \urcorner \urcorner \quad ,$$

where L is defined as in definition 4. The first three disjuncts treat
the regions L, R_1, R_3. Region R_2 is dispatched by the fourth disjunct:
Any point of R_2 must first be pushed down into region R_1 in accord with
clause (iii), and then parallel to the diagonal into his testing region
P_2 according to clause (ii). Thus N = N($\mathcal{G}$), where

$$\mathcal{G}(a,b) \equiv_{df} \ulcorner a \leq b \wedge \phi(a,b) \urcorner \vee \ulcorner b \leq a \wedge \phi(b,a) \urcorner . \quad #$$

It is easy to see how to generalize the above considerations to arbitrary n: If one wants to test a given point $\underline{m}^n$ whether it belongs
to a fanlike u.p. set N, first one has to move it orthogonally towards
the diagonal, to diminish the distances between the components; second-

ly one has to push down the point parallel to the diagonal into its
testing region. Thus for any permutation of the components, one of
which will order them by their size, one has to distinguish n cases:

Definition 5: An n-set N is <u>fanlike ultimately periodic</u> of the <u>phase</u>
f and the <u>period</u> z iff for every n-number $\underline{m}$: $\underline{m} \in N$ iff
$\underline{m} \in L =_{df} \{ \underline{\ell} \in N; \; \overset{n}{\underset{\nu=1}{\bigwedge}} \; \ell_{\nu} < n \cdot (f+z) \}$, or there is a permutation π of the
numbers $1, \ldots, n$ so that $m_{\pi(1)} \leq \ldots \leq m_{\pi(n)}$ and one of the following
two conditions is fulfilled:

(i) $f + z \leq m_1$ and $(R, m_{\pi(2)} - m_{\pi(1)} + R, \ldots, m_{\pi(n)} - m_{\pi(1)} + R) \in N$

where R is so that $f \leq R < f + z$ and $m_{\pi(1)} \equiv R \;(z)$

(ii)   there is an index $\hat{j}$, $1 \leq \hat{j} < n$, so that

   $f + z \leq m_{\pi(\hat{j}+1)} - m_{\pi(\hat{j})}$ and

   $(m_{\pi(1)}, \ldots, m_{\pi(\hat{j})}, m_{\pi(\hat{j})} + R, m_{\pi(\hat{j}+2)} - m_{\pi(\hat{j}+1)} + m_{\pi(\hat{j})} - R, \ldots, m_{\pi(n)} - m_{\pi(\hat{j}+1)} + m_{\pi(\hat{j})} - R) \in N$

where R is so that $f \leq R < f + z$ and $m_{\pi(\hat{j}+1)} \equiv m_{\pi(\hat{j})} + R \;(z)$.

   Using this definition the proof of Lemma 1 is straightforward for
the general case, too; to write down the formula of lemma 2 for the
general case is somewhat more cumbersome. We shall, however, not bother
to give these proofs, since they would not yield new insights into the
structure of fanlike u.p. sets.

Remark 2: For $n = 1$ the three properties of an n-set, to be u.p., u.p.
in layers, resp. fanlike u.p., coincide.

   Thus we conclude:

Theorem 2: Exactly the fanlike ultimately periodic sets are definable
by SC-formulae containing free individual variables. Moreover: From a
given formula f one can construct effectively the phase, the period,
and the initial portion of $N(f)$, and conversely.

   The effectiveness of theorem 2 is obvious what regards lemma 1. For
the proof of lemma 2 one has to use theorem I.2.b.2 to decide the Σ_1^{ω}-
sentences $\vartheta_{\ell}^{\omega}$, and thus to construct the set K.

   Theorem 2 shows once more the power of predicate quantifiers (cf.1.a):
In the system of Elgot-Wright[13] exactly the quasi-finite sets are definable. Here a set is called <u>quasi-finite</u> iff it is finite or has a
finite complement.

b) Quantifier elimination for SC, definable functions

In 1.b we introduced the elementary number theories P and CO of
addition and of congruence and order respectively, which are both decidable and allow effective quantifier elimination. Corollary 1.b.3a
stated that any thread definable in SC can be defined explicitly involving means only from CO, i.e. can be defined by an equivalence

(1) $(\forall t)\lceil \underline{A}(t) <-> \underline{\mathcal{C}}(t)\rceil$,

where $\mathcal{C}_i$ are CO-formulae, which are even quantifier-free.

In the preceding section we have proved a similar, but much stronger
connection between SC and CO regarding formulae with free individual
variables. For better formulation we assume now that SC contains < and
$\equiv (\mathit{w})$, $\mathit{w} = 1,2,\dots$ as primitive constants; thus CO is a subsystem of
SC. Then theorem a.2 yields

__Lemma 1:__ To any SC-formula not containing free predicate variables one
can construct effectively an equivalent CO-formula.

__Proof:__ Let $\mathit{f}(\underline{a}^{\mathit{w}})$ be given. By theorem a.2, $N(\mathit{f})$ is fanlike u.p., and
phase, period and initial portion of $N(\mathit{f})$ can be determined effectively.
Therefore $N(\mathit{f})$ can be defined by a formula of form (4) of the beginning
of section a in case $\mathit{w} = 1$, or by a formulae as in the proof of lemma
a.2 in case $\mathit{w} = 2$, or by a formula which is constructed analogously
from definition 4 in case $\mathit{w} > 2$. All these formulae are in CO, and depend only on the above constants of $N(\mathit{f})$. Since two formulae are equivalent, if they define the same set, the lemma is proved. #

Remark that in lemma 1 - as opposed to the above quoted corollary
1.b.3a - one cannot dispense with order, thus replacing CO by the elementary theory C of congruence alone. Since congruence is symmetric and
order is not, order is not definable from congruence.

Lemma 1 has as an immediate consequence

__Corollary 1:__ In CO and in SC, the same n-sets of natural numbers, sc.
the fanlike ultimately periodic sets, are definable.

Conversely, we could prove corollary 1 directly: We would have to
show by inspection that any quantifier-free CO-formula defines a fanlike
u.p. set and then to apply the quantifier elimination result for CO, and
we would have to argue as in the proof of lemma 1 to show that any fanlike u.p. set is definable in CO. Then lemma 1 would be derived from
corollary 1. - Quantifier elimination in CO together with lemma 1 yields

__Theorem 1:__ SC allows effective quantifier elimination for formulae without free predicate variables.

It is well-known (cf. e.g. Hilbert-Bernays[20], vol.I,p.378) that in the Presburger system P exactly the u.p. sets are definable. Therefore in case $w = 1$ corollary 1 holds for P instead of CO, too. But just as in section a the analogy between definable w-threads and definable w-sets did not extend from $w = 1$ to higher dimensions, we cannot replace CO by P in corollary 1 for $w > 1$. By corollary 1, any fanlike ultimately periodic w-set is definable in P, since CO is a subsystem of P (cf.section a). But the converse is by no means true. E.g. the set defined by the formula $a = 2b$ is not even ultimately periodic. Thus one has to widen this latter concept to get a description of the w-sets definable in P; we could characterize them roughly as the Boolean algebra generated by the sets which are fanlike ultimately periodic not only with respect to the diagonal, but with respect to any rational direction.

This comparison shows once more the weakness of monadic second order logic. SC allows quantification over properties of natural numbers, and thus should be expected to be stronger than any decidable elementary theory. But what regards definability of sets of natural numbers, SC is equivalent just to the simple theory CO, and is much weaker than the Presburger system P. The reason is to be found in the consideration following corollary I.2.b.4: Since one-place predicate variables together with the successor function can transmit information through a limited intervall of time only, the free individual variables of an SC-formula can be connected with each other and with the zero element only over limited distances, too. Therefore e.g. any 2-set definable in SC can show irregularities only near to the axes and to the diagonal (cf. pict.3 in section a), whereas in the remaining region the components can be shifted independently. In contrary to this, 2-sets definable in P can be continued periodically, too, but into several directions.

Theorem a.2 also yields easily a complete survey on the functions definable in SC. We will restrict ourselves to one-place functions, since such functions yield the most interesting examples in studying undecidable extensions of SC (see [35], and the introduction). (It presents no difficulties, but is more cumbersome in formulation, to extend the description to higher dimensions.)

<u>Definitions:</u> Let g be a one-place function from the natural numbers, into the natural numbers:

1) The <u>function</u> g is <u>ultimately periodic</u> (<u>u.p.</u> for short) iff there exist a <u>phase</u> ϕ and a period $\gamma > 0$ such that $g(a) = g(a+\gamma)$ for any $a \geq \phi$.

2) g is <u>essentially increasing</u> iff it exceeds ultimately any given value forever , i.e.

$$(\forall x)(\exists y)(\forall z)_y \ x \leq g(z) \ .$$

3) An w-place <u>function</u> h is <u>definable in SC</u> iff the $w+1$-set representing h is definable in SC.

  A look on pict.3 in section a shows

<u>Theorem 2:</u> Let g be a one-place function: g is definable in SC iff
there are numbers f and γ and a one-place function h   such that h is
ultimately periodic with phase f and period γ, and $(\forall t)$ $o < h(t) < 2f$,
and one of the following three conditions is fulfilled:
(1) $g(a) = h(a)-1$ for all a (g is an u.p. function)
(2) $g(a) = a+h(a)-f$ for all a (g is ultimately a periodic deformation
       of the identity function)
(3)   there is a covering M_1, M_2 of the set $\{f, \ldots, f + \gamma - 1\}$ so that

$$g(a) = \begin{cases} a+h(a)-f; & a \equiv \kappa\,(\gamma) \wedge f \leq \kappa < f+\gamma \wedge \kappa \in M_1 \\ h(a)-1; & a \equiv \kappa\,(\gamma) \wedge f \leq \kappa < f+\gamma \wedge \kappa \in M_2 \end{cases}$$

(g is a periodic combination of (1) and (2))

<u>Corollary 2:</u> Among the essentially increasing functions, exactly the
u.p. deformations of the identity function are definable in SC. We call
phase and period of the function h of theorem 2 also <u>phase</u> and <u>period</u>
of g .

<u>Corollary 3</u> (Büchi[2],[3]): Let $<$ be a wellordering on $\mathbb{N}$ of type α: $<$
is definable in SC if and only if $\alpha < \omega^2$.

<u>Proof:</u> Let $\alpha < \omega^2$, e.g. $\alpha = \omega \cdot w + w$. We use the congruence classes modu-
lo $w+1$ to represent w times the type ω and finally the type w: Let
$f(a,b)$ be the formula

$$\bigvee_{\substack{i,\gamma=0 \\ i \leq \gamma}}^{w} \lceil (a,b) \equiv (i,\gamma)(w+1) \wedge \lceil i = \gamma \to a < b \rceil \wedge$$
$$\wedge \; \lceil \gamma = w \to \bigvee_{\ell=1}^{w} b = \ell \cdot (w+1)-1 \rceil \rceil$$

Then f defines a wellordering of type α. Conversely, assume the wellordering $<$ of type α to be defined by an SC-formula f, where α is a
limit number. Let g be the successor function with respect to $<$.
Since f is an SC-formula, g clearly is definable in SC. Since g is
one-to-one, by corollary 2 it must be of the type (2) in theorem 2. Let
f and γ be the phase resp. the period of g . Then for any $a \geq f$ the difference between a and g(a) is smaller than f. This implies that the
type ω is contained at most f times in the wellordering $<$. A similar
argument holds for successor ordinals. $\#$

  Corollary 3 is proved in Büchi [2] for the system W2A (cf.3.a; for an
error in Büchi's proof see McNaughton[24]). Thus one gets another proof

for corollary 3 using theorem 3.a.6. Likewise theorem a.2 for the case
$w = 1$ is proved in Büchi[2], and the equivalence (1) of lemma a.1 is
given. The concept "fanlike ultimately periodic" and thus theorem a.2
for $w \geq 2$, seems to be new; so are theorems 1+2 of this section. Theorem
2 corrects the remark in Büchi[2], p.79, about functions definable in
W2A. For details see [35].

Corollary 3 leeds to the following question: Let $SC(\alpha)$ be like SC,
but having as domain the ordinal number α instead of ω (thus $SC(\omega) =$
SC). For what ordinals α is $SC(\alpha)$ decidable? Ersov[14] has shown that
$SC(\omega^2)$ is decidable; more generally $SC(\alpha)$ is decidable for any countable ordinal α by Büchi[5]. Our completeness proof for SC suggests that
$SC(\alpha)$ is decidable for any ordinal α for which the Ramsey theorem holds.

The investigations of this section suggest the question which sets
and relations (especially which functions) disturb the decidability of
SC when added, and which do not. We will work on this question in chap.
III.

We conclude the section by a last improvement of the DP for SC,
which is a consequence of theorem a.2. Let an SC-sentence $\mathcal{Y}$ be given
of the form $(\exists \underline{x}^w) f(\underline{x})$ or $(\forall \underline{x}^w) f(\underline{x})$. To bring $\mathcal{Y}$ into the form Σ_1^ω, one
has to drive in the initial individual quantifiers - a procedure which
will produce a lot of predicate quantifiers in general; and we know
from I.4.b what predicate quantifiers cost. Thus in most cases it will
be much easier to test the set $N(f)$ with the help of theorem a.2. Ob-
viously, $(\exists \underline{x}) f(\underline{x})$ is true iff $N(f) \neq \emptyset$, and $(\forall \underline{x}) f(\underline{x})$ is true iff
$N(f) = \mathbb{N}^w$. Since theorem a.2 is effective, we have an easy DP for such
formulae. Instead of deciding e.g. the sentence $(\forall x) x' \neq o$ as in I.4.d,
we can test the formula $a' \neq o$, which has the easy $\Sigma_{1,1}^\omega$ -form

$$(\exists PR). \neg P(o) \wedge (\forall t) \lceil P(t') <-> R(t) \rceil \wedge R(a) \; .$$

By the way, if a sentence $\mathcal{Y}$ is of the form $(\forall \underline{P}) f(\underline{P})$, it will be
better to decide $\neg \mathcal{Y}$; in case of $(\exists \underline{P}) f(\underline{P})$ it may sometimes be better in
view of corollary I.2.b.4, to decide on satisfiability of $f(\underline{A})$.

c) Standard models of SC

To look for arbitrary models of SC would exceed the frame of these lecture notes. Thus we assume as before "the" set $\mathbb{N}$ of natural numbers to be given somehow, and ask for "standard models" of SC, i.e. for the subsets of the powerset $\pi(\mathbb{N})$ of $\mathbb{N}$ which might serve as the range of interpretation of the predicate variables. The results of sections 1.b and a lead to a complete description of these standard models of SC. In fact the restriction to standard models is not essential: SC is complete, and therefore exactly the SC-sentences which are true in $\mathbb{N}$ hold in SC; thus SC is a true picture of the structure $\langle \mathbb{N}, o, ', <, \equiv (n) \rangle$.

Definition 1: Let T be a monadic second order system of number theory (see I.1.a), let $\mathcal{M}$ be a Boolean subalgebra of the power algebra $\pi(\mathbb{N})$: $(\mathbb{N}, \mathcal{M})$ is a <u>standard model of T</u> iff any sentence which holds in T becomes true if the predicate variables are interpreted as elements of $\mathcal{M}$ and the remaining signs have their usual meaning.

Let $\pi_{uper}(\mathbb{N})$ be the set of all ultimately periodic sets of natural numbers:

Theorem 1: Let $\mathcal{M}$ be any Boolean subalgebra of $\pi(\mathbb{N})$: $(\mathbb{N}, \mathcal{M})$ is a standard model of SC if and only if $\pi_{uper}(\mathbb{N})$ is contained in $\mathcal{M}$. Especially, $(\mathbb{N}, \pi_{uper}(\mathbb{N}))$ is a standard model of SC.

Proof: Let $\pi_{uper}(\mathbb{N}) \subseteq \mathcal{M}$, let $\mathcal{G}$ be any sentence which holds in SC. We want to show: $\mathcal{G}$ is true in $(\mathbb{N}, \mathcal{M})$. By theorem I.4.a.1 we may assume $\mathcal{G}$ to be of the form $(\exists \underline{P}^{n}) f(\underline{P})$ where f does not contain any predicate quantifiers. Since $\mathcal{G}$ holds in SC, $S(f)$ is not empty. By corollary 1.b.1, $S(f)$ contains an ultimately periodic thread φ. Let the components of φ be the indicators of the ultimately periodic sets $M_1, .., M_n$. Interpreting P_ν by M_ν we get $\mathcal{G}$ satisfied in $(\mathbb{N}, \mathcal{M})$. - Conversely, let $(\mathbb{N}, \mathcal{M})$ be a standard model of SC, let N be an ultimately periodic set. We want to show: $N \varepsilon \mathcal{M}$. Let φ be the indicator of N. Using corollaries 2 and 3a of 1.b we construct a formula $f(a)$ not containing predicate variables so that $N = N(f)$, and therefore

$$\mathcal{G}(A) \equiv_{df} (\forall t) \lceil A(t) <-> f(t) \rceil$$

defines φ in $(\mathbb{N}, \pi(\mathbb{N}))$. From (COMP) follows that $(\exists P) \mathcal{G}(P)$ holds in SC; therefore $(\exists P) \mathcal{G}(P)$ is true in $(\mathbb{N}, \mathcal{M})$. Since f does not contain predicate variables, $\mathcal{G}(A)$ is satisfiable in $(\mathbb{N}, \mathcal{M})$ only by φ, too. Thus $N \varepsilon \mathcal{M}$. #

Corollary 1 (Büchi[3], theorem 3): The two structures $(\mathbb{N}, \pi(\mathbb{N}))$ and $(\mathbb{N}, \pi_{uper}(\mathbb{N}))$ are equivalent with respect to monadic second order successor number theory, i.e. among the sentences expressible in this language, the same hold in both structures.

Since Büchi presents his system semantically, he speaks of a new system SC_{per} instead of the structure $(\mathbb{N}, \pi_{uper} (\mathbb{N}))$. Thus his theorem 3 is formulated for systems rather than for structures. For this reason he cannot use his definability result, but has to show directly that the same sentences hold in SC and SC_{per}.

§3. Variants of SC

The preceding section suggests to examine the structures $(\mathbb{N}, \mathcal{M})$
where $\mathcal{M}$ is a Boolean subalgebra of the power algebra $\pi(\mathbb{N})$ not containing the ultimately periodic sets. The only interesting cases seem
to be the following:
(1) $\mathcal{M}$ is the set of the finite sets, (2) $\mathcal{M}$ is the set of the quasi-
finite sets (cf. the end of 2.a). According to theorem 2.c.1 we have
to look for new systems which describe these structures. This will be
done in the next two sections. In section c we shall extend the results from the natural numbers to the integers.

a) The system SC_{fin}

Let π_{fin} ($\mathbb{N}$) be the set of all finite sets of natural numbers. In
[2], Büchi sets up semantically a system called W2A (= weak second
order arithmetic), which describes just the structure $(\mathbb{N}, \pi_{fin}$ $(\mathbb{N}))$.
W2A is like SC, but only finite threads are allowed as interpretations
of the predicate variables. Büchi shows the decidability of W2A, and
gives a rather complete description of the definable sets of both,
words and natural numbers. Sometimes we have refered already to this
description. (Clearly just the finite predicates, which can be identi-
fied with words, are definable in W2A; thus the question for the definable sets of predicates is of no interest).
According to remark 4 on p.9 of Büchi[3] another proof of the decidability of W2A follows from the decidability of SC: For any sentence
$\mathcal{G}$ of W2A let the "relativization $\mathcal{G}^{(F)}$ of $\mathcal{G}$ with respect to (F)" be
the sentence obtained from $\mathcal{G}$ by replacing any subformula $(\forall P)\, \alpha(P)$ by
$(\forall P)\lceil(\forall^\omega t)\neg P(t) \rightarrow \alpha(P)\rceil$, and analogously $(\exists P)\,\alpha(P)$ by
$(\exists P)\lceil(\forall^\omega t)\neg P(t) \wedge \alpha(P)\rceil$ (for details see end of 4.a). Clearly $\mathcal{G}$ is
true in W2A if and only if $\mathcal{G}^{(F)}$ is true in SC; in termini of 4.a:
the kernel of the relativization of W2A is straightly interpretable in
SC. Thus theorem I.4.a.2 yields a DP for W2A. - A third DP for W2A is
to be found in Elgot[11]; he calls this system L_1^1. Moreover both, Büchi
and Elgot, refer to an unpublished result of Ehrenfeucht of the same
kind.

Theorem 1 (Büchi[2], corollary 1): The system W2A is decidable.

We shall give now a syntactical version of W2A. Let SC_{fin} be the
theory arising from SC by the following modifications: (1) Change the
substitution rule (SP) into

$$(\text{SP}_{\text{fin}}) \qquad \frac{\alpha(A), \ (\forall^{\omega} t)\neg \mathcal{L}(t)}{\alpha(\mathcal{L})}$$

(2) Replace the induction axiom (I) by the induction schema (IS)

(3) Add the finiteness axiom

$\quad$ (F) $\qquad (\forall^{\omega} t)\neg A(t)$

If we interprete SC_{fin} like W2A, it is clear that these three modifications are necessary: (SP) is not compatible with W2A, (I) is useless in W2A, and (F), though true in W2A, is not derivable from the remaining axioms, since it is not derivable in SC. Conversely, the modifications are also sufficient: The DP for W2A of Büchi[2] uses nearly no means than the theory of automata. To eliminate these means in SC we (a) derived the "finiteness considerations", which are typical for the automata theory (lemmata 1 and 2 in I.5.a), and (b) proved the recursion theorem I.1.b.1. The former is done easily in SC_{fin}, too. What regards (b) it suffices to derive the recursion theorem in the restricted form of lemma I.1.b.2:

$$(\exists P)(\forall x)_{o}^{a}\lceil P(x) <\text{-}> \mathcal{L}(x, P(t); t < x)\rceil \quad .$$

This derivation is possible in SC_{fin}, too. Thus we conclude:

<u>Theorem 2</u>: The theory SC_{fin} is complete.

Once more let us consider the modifications which led us from the axioms of SC to those of SC_{fin}. Modification (2) is clearly inessential: We might likewise have built up SC with (IS) instead of (I). Further, just as (SP) is equivalent to (COMP) within SC, (SP_{fin}) is equivalent within SC_{fin} to the modified comprehension principle

$$(\text{COMP}_{\text{fin}}) \qquad (\forall^{\omega} t)\neg \alpha(t) \text{-}> (\exists P)(\forall t)\lceil P(t) <\text{-}> \alpha(t)\rceil \quad .$$

Thus the axiom system of SC_{fin} coincides with that of SC with the exception of the two axioms (F) and $(\text{COMP}_{\text{fin}})$, the relativizations of which with respect to (F) are derivable in SC. Therefore theorem 2 should also be obtained by translating the derivations of SC into derivations of W2A and then using the completeness of SC. We did not, however, succeed in showing this.

In spite of the fact that W2A is rather weak compared to SC, we shall show now that both systems are of equal strength with regard to definability.

<u>Theorem 3</u> (Büchi[2], theorem 8): Exactly the word sets definable by finite automata are definable in W2A.

__Theorem 4__ (Büchi[2], theorem 3): Any formula $f(\underline{A}^w, \underline{a}^m)$ is equivalent in W2A to a formula of the form

$$(\exists \underline{P}) . \alpha[\underline{P}(o)] \wedge (\forall t) \mathcal{L}[\underline{A}(t), \underline{P}(t), \underline{P}(t')] \wedge \bigwedge_{v=1}^{m} f_v(a_v)$$

__Theorem 5__ (For $w = 1$: Büchi[2], corollary 2): Exactly the fanlike ultimately periodic w-sets are definable in W2A.

For the proof of theorems 3 and 4 see Büchi[2]. To prove theorem 5 note first that $<$ and $\equiv(w)$, $w = 2, 3, \ldots$ are definable in W2A similarly as in SC; indeed, the definition of order can be taken unchanged, whereas the definition of congruence carries over if we replace P by ¬P in the defining formula in I.1.a. Thus CO is a subsystem of W2A, and by corollary 2.b.1 any fanlike ultimately periodic w-set is definable in W2A. Conversely, let $f(a,b)$ be any formula of W2A; f may be assumed to be in the form of theorem 4. Modify the proof of lemma 2.a.1 by deleting anywhere the subformula $(\exists^w t) \mathcal{L}[P(t)]$. Then we get a proof that f defines a fanlike u.p. set in W2A. Since lemma 2.a.1 extends to arbitrary w as in SC, the theorem is proved.

In fact, it was corollary 2 of Büchi[2] (corresponding to our remark 2.a.1) and the equivalence on p.79,1.c., (corresponding to the equivalence (1) in the proof of our lemma 2.a.1) which led us to theorem 2.a.2 and to the definition of "fanlike ultimately periodic".

Since theorems 3-5 are effective, we get
__Theorem 6__ (Büchi[3], remark 4,p.7): The same sets of words and the same w-sets are definable in SC and in W2A. Moreover, from a formula in one system one can get effectively a formula in the other system which defines the same set.

The part of theorem 6 which is related to w-sets answers a question raised in R.M. Robinson[33]. It shows that, from a mathematical point of view, both systems are of equal strength: the same properties of natural numbers are derivable, since the same properties are definable.
We could prove this part of theorem 6 without using theorem 5: Let $\mathcal{G}(a,b)$ be the formula equivalent to $f(a,b)$ in (1) of the proof of lemma 2.a.1. Since $\mathcal{G}$ contains only restricted predicate quantifiers, $\mathcal{G}(a,b)$ is true in SC if and only if it is true in W2A. Thus $\mathcal{G}$ is a formula defining the same 2-set in W2A as the formula f in SC. Interchanging SC and W2A we get the converse result this time using the version of the quoted lemma 1 modified for W2A as above. - The above proof of theorem 6 yields

__Theorem 7:__ W2A allows effective quantifier elimination for formulae not containing free predicate variables. Indeed, to any such formula one can construct an equivalent formula in CO.

b) The system SC$_\text{qfin}$

As W2A is got from SC by restriction to finite predicates, we get
the system L_1^3 of Elgot[11], section 5.15, from SC by admitting only
quasi-finite sets (cf. the beginning of §3) as interpretations of the
predicate variables. Analogously we get a syntactical formulation of L_1^3
from SC by changing (I) into (IS), (SP) into

$$(\text{SP}_\text{qfin}) \quad \frac{\alpha(A), (\forall^\omega t)\, \mathcal{L}(t) \vee (\forall^\omega t)\neg\, \mathcal{L}(t)}{\alpha(\mathcal{L})}$$

and adding

$$(\text{QF}) \qquad (\forall^\omega t)A(t) \vee (\forall^\omega t)\neg A(t)$$

Let us interprete this theory like L_1^3 and call the resulting system
SC$_\text{qfin}$.

__Theorem 1__ (Elgot[11],5.15, corollary 1): The system L_1^3 is decidable.

__Proof:__ See Elgot. The DP of Elgot uses the theory of automata and the
concept of regular sets (1.a). - A second DP follows from theorem
I.4.a.2 by relativizing the L_1^3-sentences with respect to (QF) (analogously to the case of W2A). - A third DP is got by splitting the predicate quantifiers: Let $\mathcal{G}$ be a sentence in L_1^3. Replace any subformula
$(\exists P)\, \alpha(P)$ by $(\exists P)\lceil \alpha(P) \vee \alpha(\neg P)\rceil$ and analogously $(\forall P)\, \alpha(P)$ by
$(\forall P)\lceil \alpha(P) \vee \alpha(\neg P)\rceil$. Clearly the resulting formula $\tilde{\mathcal{G}}$ is true in W2A
if and only if $\mathcal{G}$ is true in L_1^3. Thus the DP for W2A yields a DP for
L_1^3. #

__Theorem 2:__ The theory SC$_\text{qfin}$ is complete.

__Proof:__ As in the case of W2A, a careful analysis of the DP shows that
the portion of the theory of regular sets and finite automata which is
used for the DP is expressible in the language and derivable from the
axioms. - In this case we get the completeness result also easily from
the one of SC$_\text{fin}$ (theorem a.2): Let $\mathcal{G}$ be any sentence of the language
considered, let $\tilde{\mathcal{G}}$ be the sentence resulting from $\mathcal{G}$ by splitting the
predicate quantifiers as in the proof of theorem 1. Since W2A is com-
plete, $\tilde{\mathcal{G}}$ or $\neg\tilde{\mathcal{G}}$ is derivable in W2A. Let e.g. $\tilde{\mathcal{G}}$ be derivable. Rela-
tivize any formula in the derivation with respect to (F) (the relati-
vized formula to a formula $f(\underline{A}^n)$ containing free predicate variables
is $\bigwedge_{\nu=1}^{n} (\forall^\omega t)\neg A_\nu(t) \rightarrow f(\underline{A})$). All the relativized axioms and rules of
SC$_\text{fin}$ are easily derivable in SC$_\text{qfin}$. Thus the relativization of the
SC$_\text{fin}$-derivation of $\tilde{\mathcal{G}}$ yields a derivation of $\tilde{\mathcal{G}}^{(F)}$ in SC$_\text{qfin}$, which is
extended using the axiom (QF) to a derivation of $\mathcal{G}$. Just so one gets
a derivation of $\neg\mathcal{G}$ in the other case; therefore SC$_\text{qfin}$ is complete. #

__Theorem 3:__ The same w-sets and the same sets of words are definable in SC and L_1^3. Moreover, one can get effectively the corresponding formulae from one system for the other.

__Proof:__ Let N be a fanlike u.p. w-set, let $f(\underline{a}^w)$ be a formula defining N in W2A (theorem a.5). Clearly, for any $\underline{w}^w$ $f(\underline{w})$ is true in W2A if and only if the relativization $(f(\underline{w}))^{(F)}$ is true in L_1^3. Thus N is definable in L_1^3 by $f^{(F)}$. Conversely, let $f(\underline{a}^w)$ be any formula of this language. Analogously, for any $\underline{w}^w$ $f(\underline{w})$ is true in L_1^3 if and only if the relativization $(f(\underline{w}))^{(QF)}$ is true in SC. By theorem 2.a.2 f defines a fanlike ultimately periodic set in L_1^3. - An analogous argument gives the corresponding result for sets of words. That just the regular sets are definable in L_1^3 is shown in Elgot[11], too. #

__Theorem 4:__ L_1^3 allows effective quantifier elimination for formulae not containing free predicate variables. Indeed, to any such formula one can construct an equivalent one in the subsystem CO.

c) Translation to the integers

By a few modifications one gets from the DP for SC a DP for the corresponding system for the integers instead of the natural numbers; an analogous result holds for W2A and L_1^3.

Thus let $\mathbb{Z}$ be the set of the integers, let $SC^{\mathbb{Z}}$ be the system with the same signs as SC, reinterpreted in $\mathbb{Z}$ in the obvious way; e.g. arbitrary subsets of $\mathbb{Z}$ are admitted as interpretations of the predicate variables. The logical rules and axioms are the same for $SC^{\mathbb{Z}}$ as for SC, too. As non-logical axioms we choose the following:

(A1) $a' = b' \rightarrow a = b$

(A3) $(\exists x)\, a = x'$

(A5) $o \neq o'$

(I^+) $A(o) \wedge (\forall t)_o \lceil A(t) \rightarrow A(t') \rceil \rightarrow (\forall t)_o A(t)$

(I^-) $A(o) \wedge (\forall t)^o \lceil A(t') \rightarrow A(t) \rceil \rightarrow (\forall t)^{o'} A(t)$

(I^+) and (I^-) are the induction axioms for the non-negative resp. the non-positive integers. As in SC one gets from it the corresponding induction schemata. With its help one shows easily that the axioms describe the set $\mathbb{Z}$ adequately, e.g. from (A5) follows that for any w $a + w \neq a$.

Here $a < b$ and $a + w$ are defined as in SC (I.1.a). We take over to $SC^{\mathbb{Z}}$ also the remaining definitions from I.1.a, and supplement it by the following ones:

1) $a = {}'b \, <->_{df} \, a' = b$

2) $(\exists t)^a\, \alpha(t) \, <->_{df} \, (\exists t)\lceil t < a \wedge \alpha(t) \rceil$

3) $(\forall t)^a\, \alpha(t) \, <->_{df} \, (\forall t)\lceil t < a \rightarrow \alpha(t) \rceil$

4) $(\exists_w t)\, \alpha(t) \, <->_{df} \, (\forall x)(\exists y)^x\, \alpha(y)$

5) $(\forall_w t)\, \alpha(t) \, <->_{df} \, (\exists x)(\forall y)^x\, \alpha(y)$

6) $(\exists P)_w\, \alpha(P) \, <->_{df} \, (\exists P)\lceil (\exists_w t)\, P(t) \wedge \alpha(P) \rceil$

Further we retain the definition of Σ^o-formula, and confront it with the following one:

$\underline{\Sigma}_o$ is the set of all formulae $\mathfrak{f}(\underline{A}, a, b)$ of the form

$$(\exists \underline{P}).\, \alpha\lfloor \underline{P}(a) \rfloor \wedge (\forall t)^{a'}_{b}\, \mathfrak{J}[\underline{A}(t), \underline{P}(t), \underline{P}({}'t)] \wedge \mathfrak{L}\lfloor \underline{P}(b) \rfloor$$

Thus $\underline{\Sigma}_o$-formulae correspond to non-deterministic finite automata "with reversed time", but have clearly the same properties as Σ^o-formulae.

Especially we can define analogously Σ_o^R-formulae, and prove the theorems of I.2.b+c and I.3.a for Σ_o-formulae instead of Σ^o-formulae. – Further we can derive the recursion theorem I.1.b.1 and the theorem of Ramsey (I.1.c.3), first restricted to the natural numbers and secondly, in adequate formulation, restricted to the non-positive integers.

Since we want to have Σ_N^ω and especially Σ_1^ω as normal forms for $SC^{\mathbb{Z}}$-formulae, we cannot simply distinguish between the positive and the negative part of Σ_N^ω-formulae, as we did with the induction axiom and with Σ^o-formulae. We define:

$\underline{\Sigma_1^{\omega^*+\omega}}$ is the set of all formulae of the form

$$(\exists\underline{P}).\mathcal{A}[\underline{P}(o)] \wedge (\forall t)_o \mathcal{L}_1[\underline{A}(t),\underline{P}(t),\underline{P}(t')] \wedge$$
$$\wedge (\forall t)^{o'} \mathcal{L}_2[\underline{A}(t),\underline{P}(t),\underline{P}('t)] \wedge (\exists^\omega t) \mathfrak{L}_1[\underline{P}(t)] \wedge (\exists_\omega t) \mathfrak{L}_2[\underline{P}(t)]$$

Analogously we define the set $\prod_1^{\omega^*+\omega}$. From this we define $\Sigma_N^{\omega^*+\omega}$ and $\prod_N^{\omega^*+\omega}$ as in SC.

Now we are able to translate the SC-DP into $SC^{\mathbb{Z}}$. Clearly $\Sigma_1^{\omega^*+\omega}$-sentences are decidable by exactly the same methods as Σ_1^ω-sentences. To show that $\Sigma_1^{\omega^*+\omega}$ is a normal form in $SC^{\mathbb{Z}}$ we remark first that the theorems 1+2 of I.3.b hold in $SC^{\mathbb{Z}}$, too; the proofs are easily modified. Now look at the proof of theorem I.3.b.3: Let f be a $\Sigma_1^{\omega^*+\omega}$-formula. Construct as in SC formulae ϑ_ν, φ_ν, and $g_{\nu,\gamma}$ for the "positive part" of f and analogously ϑ_ν', φ_ν', $g_{\nu,\gamma}'$ for the "negative part" (ϑ_ν' and φ_ν' are Σ_o-formulae, $g_{\nu,\gamma}'$ are reversed Σ_1^ω-formulae – not $\Sigma_1^{\omega^*+\omega}$-formulae – coming from the modified theorem of Ramsey). Then by obvious modifications we get

$$g_{\nu,\gamma}'(\underline{A}) \wedge g_{R,e}(\underline{A}) \wedge g_{\nu,\gamma}'(\underline{B}) \wedge g_{R,e}(\underline{B}) .\text{-}>. f(\underline{A}) <\text{-}> f(\underline{B})$$

as analogue to lemma I.3.b.2. From this we get again

$$\neg f <\text{-}> \bigvee_{(\nu,\gamma,R,e)\in M} \lceil g_{\nu,\gamma}' \wedge g_{R,e} \rceil$$

where $M =_{df} \{(\nu,\gamma,R,e); \neg(\exists\underline{P}).f(\underline{P}) \wedge g_{\nu,\gamma}'(\underline{P}) \wedge g_{R,e}(\underline{P})\}$.

Theorem I.3.a.1 is easily modified to show that $g_{\nu,\gamma}' \wedge g_{R,e}$ are $\Sigma_1^{\omega^*+\omega}$-formulae. It remains to extend theorem I.1.d.1 to get the normal form $\Sigma_N^{\omega^*+\omega}$: The proof remains unchanged up to the point where one has brought $Q_2 f_1$ into the Behmann disjunctive normal form. Within the resulting formula split up the individual quantifiers into positive and negative part, i.e. replace each $(\forall t) g_\nu(t)$ by $(\forall t)_o g_\nu(t) \wedge (\forall t)^{o'} g_\nu(t)$ and analogously $(\exists t) \varphi_{\nu,\gamma}(t)$ by $(\exists t)_o \varphi_{\nu,\gamma}(t) \vee (\exists t)^{o'} \varphi_{\nu,\gamma}(t)$. Restoring the disjunctive normal form one gets disjuncts of the form

$$\mathcal{A}_\nu(o) \wedge (\forall t)_o \mathcal{L}_\nu(t) \wedge (\forall t)^{o'} \mathcal{L}_\nu'(t) \wedge \bigwedge_{\gamma\in L_\nu} (\exists t)_o \mathfrak{L}_{\nu,\gamma}(t) \wedge \bigwedge_{\gamma\in M_\nu} (\exists t)^{o'} \mathfrak{L}_{\nu,\gamma}'(t) .$$

To these formulae we apply, instead of remark 5 in I.1.a, the following equivalence

$$\bigwedge_{j \in L} (\exists t)_o H_j(t) \wedge \bigwedge_{j \in M} (\exists t)^{o'} H_j(t) \; :\longleftrightarrow:$$

$$:\longleftrightarrow: (\exists \underline{R}^w). \bigwedge_{\nu=1}^{w} \neg R_\nu(o) \wedge (\forall t)_o \left[\underline{R}(t') \longleftrightarrow \underline{\vartheta}(t) \right] \wedge$$

$$\wedge (\exists t)_o \bigwedge_{j \in L} R_\nu(t) \wedge (\forall t)^{o'} \left[\underline{R}('t) \longleftrightarrow \underline{\vartheta}(t) \right] \wedge (\exists t)^{o'} \bigwedge_{j \in M} R_j(t)$$

where w is the cardinality of $L \cup M$, $L \cap M = \emptyset$, and

$$\vartheta_j(a) \equiv_{df} R_j(a) \vee H_j(a).$$

The rest of the proof remains unchanged, and yields thus a Σ_1^{w+w}-formula.

This completes the translation of the SC-DP into $SC^{\mathbb{Z}}$:

Theorem 1: The system $SC^{\mathbb{Z}}$ is decidable and complete.

Wholly analogously as we have got $SC^{\mathbb{Z}}$ from SC, we define the systems L_1^2 and L_1^4 from W2A resp. L_1^3; we call the corresponding axiomatic systems $SC^{\mathbb{Z}}_{fin}$ resp. $SC^{\mathbb{Z}}_{qfin}$. DPs for L_1^2 and L_1^4 are to be found in Elgot [11], 5.12 and 5.16. Clearly one could get it also from the DP for $SC^{\mathbb{Z}}$, analogous to sections a and b. We conclude:

Theorem 2: $SC^{\mathbb{Z}}_{fin}$ and $SC^{\mathbb{Z}}_{qfin}$ are decidable and complete.

B I B L I O G R A P H Y

[1] <u>Behmann</u>, Heinrich, "Beiträge zur Algebra der Logik, insbesondere zum Entscheidungsproblem", Math.Ann. 86(1922), 163-229

[2] <u>Büchi</u>, J.Richard, "Weak second-order arithmetic and finite automata", Z.Math.Logik Grundl.Math. 6(1960), 66-92

[3] -"- "On a decision method in restricted second-order arithmetic", in"Logic Meth.Phil.Sc.,Proc. 1960 Stanford Int.Congr.", Stanford 1962, 1-11

[4] -"- "Transfinite automata recursions and weak second order theory of ordinals", in "Logic Meth.Phil.Sc.,Proc. 1964 Jerusalem Int.Congr.", Jerusalem 1965, 3-23

[5] -"- "Decision methods in the theory of ordinals", Bulletin AMS 71(1965), 767-770

[6] <u>Büchi</u>, J.Richard, <u>Landweber</u>, Lawrence H., "Definability in the Monadic Second-Order Theory of Successor", Purdue University Report CSD TR 15, Sept. 1967

[7] -"- "Solving Sequential Conditions by Finite State Strategies", Purdue University Report CSD TR 14, Sept. 1967

[8] <u>Church</u>, Alonzo, "Application of recursive arithmetic to the problem of circuit synthesis", in "Summer Inst.Symb.Logic Cornell Univ.1957", ^{2}1960, 3-50

[9] -"- "Application of Recursive Arithmetic in the Theory of Computing and Automata", in "Adv.Theory Log.Design Dig.Computers, Summer Course Michigan Univ.", 1959

[10] -"- "Logic, Arithmetic and Automata", Int.Congr.Math.Stockholm 1962, 23-35

[11] <u>Elgot</u>, Calvin C., "Decision problems of finite automata design and related arithmetics", Trans.AMS 98(1961), 21-51; 103(1962), 558-559

[12] <u>Elgot</u>, Calvin C., <u>Rabin</u>, Michael O., "Decidability and undecidability of extensions of second (first) order theories of (generalized) successor", J.Symb.Logic 31(1966), 169-181

[13] <u>Elgot</u>, Calvin C., <u>Wright</u>, Jesse B., "Quantifier elimination in a problem of logical design", Mich.Math.J. 6(1959), 65-69

[14] <u>Ershov</u>, Yu.L., "Decidability of certain non-elementary theories" (Russ.), Algebra i Logica 3(1964), 45-47

[15] Ershov, Yu.L., Lavrov, I.A., Taimanov, A.D., Taitslin, M.A.,
        "Elementary theories" (Russ.), Usp.mat.nauk 2o(1965), 37-1o8;
        Engl.transl. in Russ.Math.Surv. 2o(1965), 35-1o6

[16] Gurevič, Yu.Š., "Elementary properties of ordered Abelian groups"
        (russ.), Algebra i Logica 3(1964), 5-39; Engl.transl. in
        AMS Translations 46(1965), 165-192

[17] Hasenjaeger, Gisbert, "Über ω-Unvollständigkeit in der Peano-
        Arithmetik", J.Symb.Logic 17(1952), 81-97

[18] Henkin, Leon, "Banishing the rule of substitution for functional
        variables", J.Symb.Logic 18(1953), 2o1-2o8

[19] Hermes, Hans, "Aufzählbarkeit, Entscheidbarkeit, Berechenbarkeit",
        Berlin-Göttingen-Heidelberg 1961

[2o] Hilbert, David, Bernays, Paul, "Grundlagen der Mathematik. I+II",
        Berlin-Heidelberg-New York ^{2}1968+1969

[21] Kleene, S.C., "Representation of events in nerve nets and finite
        automata", RAND memorandum Dec.1951

[22] McNaughton, Robert, "Some formal relative consistency proofs",
        J.Symb.Logic 18(1953), 136-144

[23] -"-   "The theory of automata, a survey", Adv.Comp. 2(1961),
        379-421

[24] -"-   Review to [2] and [3], J.Symb.Logic 28(1963), 1oo-1o2

[25] Myhill, John R., "Finite automata and representation of events",
        WADC Report TR 57-627 "Fundamental Concepts in the Theory of
        Systems" (Oct.1957), 112-137

[26] Presburger, M., "Über die Vollständigkeit eines gewissen Systems
        der Arithmetik ganzer Zahlen ...", Spraw.I Kongr.Mat.Slov.
        (Warschau 1929), 92-1o1, 395

[27] Putnam, Hilary, "Decidability and essential undecidability",
        J.Symb.Logic 22(1957), 39-54

[28] Rabin, Michael O., "Decidability of second-order theories and
        automata on infinit trees",
        IBM Research Report RC-2o12, Febr.13, 1968

[29] -"-   "Weakly definable relations and special automata",
        The Hebrew University, Technical Report No.32, Jerusalem,
        June 1969

[30] Rabin, Michael, Scott, Dana, "Finite automata and their decision
        problems", IBM J.Research Dev. 3(1959), 114-125

[31] Ramsey, F.P., "On a problem of formal logic",
    Proc.London Math.Soc. 3o(1929-3o), 264-286

[32] Robinson, Julia, "Definability and decision problems in arithmetic",
    J.Symb.Logic 14(1949), 98-114

[33] Robinson,Raphael M., "Restricted set-theoretical definitions in
    arithmetic", Proc. AMS 9(1958), 238-242

[34] Scholz, Heinrich, Hasenjaeger, Gisbert, "Grundzüge der mathema-
    tischen Logik", Berlin-Göttingen-Heidelberg 1961

[35] Siefkes, Dirk, "Decidable and undecidable extensions of one-place
    second order successor arithmetic" (abstract),
    J.Symb.Logic 33(1968), 494

[36] "-" "Recursion theory and the theorem of Ramsey in one-place
    second order successor arithmetic", in "Contributions to
    Mathematical Logic", Proc.1966 Hannover Coll. (ed.K.Schütte),
    Amsterdam 1968, 237-254

[37] -"- "Decidable extensions of monadic second order successor
    arithmetic", to appear in "Formale Sprachen und Automaten-
    theorie", Tagung Oberwolfach, October 1969, Bibl.Inst.
    Mannheim

[38] Skolem, Thoralf, "Über einige Satzfunktionen in der Arithmetik",
    Skrifter Norske Vid.Akad.Oslo I.Klasse 193o,no.7, Oslo 1931

[39] Specker, E.P., Hodes, Louis, "Elimination of quantifiers and the
    length of formulae", Notices AMS 12(1965), 242

[4o] Szmielew, Wanda, "Elementary properties of Abelian groups",
    Fund.Math. 41(1955), 2o3-271

[41] Tarski, Alfred, Mostowski, Andrzej, Robinson, Raphael M.,
    "Undecidable theories", Amsterdam 1953

[42] Trahtenbrot, B.A., "Certain constructions in the logic of one-
    place predicates" (Russ.), Dokl.Akad.Nauk SSSR 138(1961),
    32o-321; Engl.transl. in Sov.Math. 2(1961), 623-625

[43] Trahtenbrot, B.A., "Finite automata and the logic of one-place
    predicates" (Russ.), Sib.Mat.Z. 3(1962), 1o3-131;
    Engl.transl. in AMS Transl. 59(1966), 23-55

List of Symbols and Notations

a) Variables used in object languages and in the metalanguage

$a,\ldots,e$	Individual (free)	3
t,x,y,z	" (bound)	3
$A,\ldots,E,G,H$	Predicate (free)	3
P,Q,R,S	" (bound)	3
$\underline{a}^w,\underline{a},\ldots$	Tuples	3
$\mathfrak{f},\ldots,\eta$	Natural numbers	3,4
$\underline{\mathfrak{f}}^w,\underline{\mathfrak{f}},\ldots$	Tuples	3,104
$\overline{\underline{\mathfrak{f}}}^w,\overline{\underline{\mathfrak{f}}},\ldots$	Constant tuples	104
$I,\ldots,N$	Sets of natural numbers	4
α	Terms	4
$\alpha,\mathscr{L},\ldots$	Formulae	4
$\underline{\alpha}^w,\underline{\alpha},\ldots$	Tuples	9
$X,Y,Z,X^w,\ldots$	States	26
u,v,w	Words	29
U,V,W	Sets of words	29
φ,χ,γ	Threads	26
Φ,X,Ψ	Sets of Threads	26
γ	Finite automata	87
S,T	Theories	115

All variables may be indexed.

b) Constants, special symbols

$=,o,',<,+w,\equiv(w)$	Number theoretic symbols	3,6,7,122
$\wedge,\vee,\neg,->,<->,=,\forall,\exists,T,F$	Logical symbols	3,5
$\lceil,\rceil,\lfloor,\rfloor,\ldots$	Brackets and dots for bracketing of	3
$\cdot,:,:\cdot,\ldots$	formulae, for quantifiers and argu-	3
$(,),[,]$	ments of predicate variables and formulae	3,4
$(\exists t)_a^b,(\forall t)_a^b,(\exists t)_a,(\forall t)_a,$ $(\exists^\omega t),(\forall^\omega t),(\exists P)^\omega$	Specialized quantifiers	6
$(\exists t)^a,(\forall t)^a,(\exists_\omega t),(\forall_\omega t),$ $(\exists P)_\omega$	"	122
$\underset{a}{\overset{b}{=}},\underset{a}{\overset{b}{\gtrless}},\underset{a}{\overset{b}{\lessgtr}}$	Relations on predicates	76
$<$	Ordering on the set of states	76
$<_M$	" on a set M	78

f) Theories and systems

Offsetdruck: Julius Beltz, Weinheim/Bergstr.

Lecture Notes in Mathematics

Bitte wenden / Continued

Vol. 72: The Syntax and Semantics of Infinitary Languages. Edited by J. Barwise. IV, 268 pages. 1968. DM 18,– / $ 5.00

Vol. 73: P. E. Conner, Lectures on the Action of a Finite Group. IV, 123 pages. 1968. DM 10,– / $ 2.80

Vol. 74: A. Fröhlich, Formal Groups. IV, 140 pages. 1968. DM 12,– / $ 3.30

Vol. 75: G. Lumer, Algèbres de fonctions et espaces de Hardy. VI, 80 pages. 1968. DM 8,– / $ 2.20

Vol. 76: R. G. Swan, Algebraic K-Theory. IV, 262 pages. 1968. DM 18,– / $ 5.00

Vol. 77: P.-A. Meyer, Processus de Markov: la frontière de Martin. IV, 123 pages. 1968. DM 10,– / $ 2.80

Vol. 78: H. Herrlich, Topologische Reflexionen und Coreflexionen. XVI, 166 Seiten. 1968. DM 12,– / $ 3.30

Vol. 79: A. Grothendieck, Catégories Cofibrées Additives et Complexe Cotangent Relatif. IV, 167 pages. 1968. DM 12,– / $ 3.30

Vol. 80: Seminar on Triples and Categorical Homology Theory. Edited by B. Eckmann. IV, 398 pages. 1969. DM 20,– / $ 5.50

Vol. 81: J.-P. Eckmann et M. Guenin, Méthodes Algébriques en Mécanique Statistique. VI, 131 pages. 1969. DM 12,– / $ 3.30

Vol. 82: J. Wloka, Grundräume und verallgemeinerte Funktionen. VIII, 131 Seiten. 1969. DM 12,– / $ 3.30

Vol. 83: O. Zariski, An Introduction to the Theory of Algebraic Surfaces. IV, 100 pages. 1969. DM 8,– / $ 2.20

Vol. 84: H. Lüneburg, Transitive Erweiterungen endlicher Permutationsgruppen. IV, 119 Seiten. 1969. DM 10.– / $ 2.80

Vol. 85: P. Cartier et D. Foata, Problèmes combinatoires de commutation et réarrangements. IV, 88 pages. 1969. DM 8,– / $ 2.20

Vol. 86: Category Theory, Homology Theory and their Applications I. Edited by P. Hilton. VI, 216 pages. 1969. DM 16,– / $ 4.40

Vol. 87: M. Tierney, Categorical Constructions in Stable Homotopy Theory. IV, 65 pages. 1969. DM 6,– / $ 1.70

Vol. 88: Séminaire de Probabilités III. IV, 229 pages. 1969. DM 18,– / $ 5.00

Vol. 89: Probability and Information Theory. Edited by M. Behara, K. Krickeberg and J. Wolfowitz. IV, 256 pages. 1969. DM 18,–/ $ 5.00

Vol. 90: N. P. Bhatia and O. Hajek, Local Semi-Dynamical Systems. II, 157 pages. 1969. DM 14,– / $ 3.90

Vol. 91: N. N. Janenko, Die Zwischenschrittmethode zur Lösung mehrdimensionaler Probleme der mathematischen Physik. VIII, 194 Seiten. 1969. DM 16,80 / $ 4.70

Vol. 92: Category Theory, Homology Theory and their Applications II. Edited by P. Hilton. V, 308 pages. 1969. DM 20,– / $ 5.50

Vol. 93: K. R. Parthasarathy, Multipliers on Locally Compact Groups. III, 54 pages. 1969. DM 5,60 / $ 1.60

Vol. 94: M. Machover and J. Hirschfeld, Lectures on Non-Standard Analysis. VI, 79 pages. 1969. DM 6,– / $ 1.70

Vol. 95: A. S. Troelstra, Principles of Intuitionism. II, 111 pages. 1969. DM 10,– / $ 2.80

Vol. 96: H.-B. Brinkmann und D. Puppe, Abelsche und exakte Kategorien, Korrespondenzen. V, 141 Seiten. 1969. DM 10,– / $ 2.80

Vol. 97: S. O. Chase and M. E. Sweedler, Hopf Algebras and Galois theory. II, 133 pages. 1969. DM 10,– / $ 2.80

Vol. 98: M. Heins, Hardy Classes on Riemann Surfaces. III, 106 pages. 1969. DM 10,– / $ 2.80

Vol. 99: Category Theory, Homology Theory and their Applications III. Edited by P. Hilton. IV, 489 pages. 1969. DM 24,–/ $ 6.60

Vol. 100: M. Artin and B. Mazur, Etale Homotopy. II, 196 Seiten. 1969. DM 12,– / $ 3.30

Vol. 101: G. P. Szegö et G. Treccani, Semigruppi di Trasformazioni Multivoche. VI, 177 pages. 1969. DM 14,–/ $ 3.90

Vol. 102: F. Stummel, Rand- und Eigenwertaufgaben in Sobolewschen Räumen. VIII, 386 Seiten. 1969. DM 20,– / $ 5.50

Vol. 103: Lectures in Modern Analysis and Applications I. Edited by C. T. Taam. VII, 162 pages. 1969. DM 12,– / $ 3.30

Vol. 104: G. H. Pimbley, Jr., Eigenfunction Branches of Nonlinear Operators and their Bifurcations. II, 128 pages. 1969. DM 10,–/ $ 2.80

Vol. 105: R. Larsen, The Multiplier Problem. VII, 284 pages. 1969. DM 18,– / $ 5.00

Vol. 106: Reports of the Midwest Category Seminar III. Edited by S. Mac Lane. III, 247 pages. 1969. DM 16,– / $ 4.40

Vol. 107: A. Peyerimhoff, Lectures on Summability. III, 111 pages. 1969. DM 8,–/ $ 2.20

Vol. 108: Algebraic K-Theory and its Geometric Applications. Edited by R. M. F. Moss and C. B. Thomas. IV, 86 pages. 1969. DM 6,–/ $ 1.70

Vol. 109: Conference on the Numerical Solution of Differential Equations. Edited by J. Ll. Morris. VI, 275 pages. 1969. DM 18,– / $ 5.00

Vol. 110: The Many Facets of Graph Theory. Edited by G. Chartrand and S. F. Kapoor. VIII, 290 pages. 1969. DM 18,– / $ 5.00

Vol. 111: K. H. Mayer, Relationen zwischen charakteristischen Zahlen. III, 99 Seiten. 1969. DM 8,– / $ 2.20

Vol. 112: Colloquium on Methods of Optimization. Edited by N. N. Moiseev. IV, 293 pages. 1970. DM 18,–/ $ 5.00

Vol. 113: R. Wille, Kongruenzklassengeometrien. III, 99 Seiten. 1970. DM 8,– / $ 2.20

Vol. 114: H. Jacquet and R. P. Langlands, Automorphic Forms on GL (2). VII, 548 pages. 1970. DM 24,– / $ 6.60

Vol. 115: K. H. Roggenkamp and V. Huber-Dyson, Lattices over Orders I. XIX, 290 pages. 1970. DM 18,– / $ 5.00

Vol. 116: Séminaire Pierre Lelong (Analyse) Année 1969. IV, 195 pages. 1970. DM 14,– / $ 3.90

Vol. 117: Y. Meyer, Nombres de Pisot, Nombres de Salem et Analyse Harmonique. 63 pages. 1970. DM 6.– / $ 1.70

Vol. 118: Proceedings of the 15th Scandinavian Congress, Oslo 1968. Edited by K. E. Aubert and W. Ljunggren. IV, 162 pages. 1970. DM 12,– / $ 3.30

Vol. 119: M. Raynaud, Faisceaux amples sur les schémas en groupes et les espaces homogènes. III, 219 pages. 1970. DM 14,– / $ 3.90

Vol. 120: D. Siefkes, Büchi's Monadic Second Order Successor Arithmetic. XII, 130 Seiten. 1970. DM 12,–